Studies in Applied Philosophy, Epistemology and Rational Ethics

Volume 36

About this Series

Studies in Applied Philosophy, Epistemology and Rational Ethics (SAPERE) publishes new developments and advances in all the fields of philosophy, epistemology, and ethics, bringing them together with a cluster of scientific disciplines and technological outcomes: from computer science to life sciences, from economics, law, and education to engineering, logic, and mathematics, from medicine to physics, human sciences, and politics. It aims at covering all the challenging philosophical and ethical themes of contemporary society, making them appropriately applicable to contemporary theoretical, methodological, and practical problems, impasses, controversies, and conflicts. The series includes monographs, lecture notes, selected contributions from specialized conferences and workshops as well as selected Ph.D. theses.

Advisory Board

More information about this series at http://www.springer.com/series/10087

Cameron Shelley

Design and Society: Social Issues in Technological Design

Springer

Cameron Shelley
Centre for Society, Technology and Values
University of Waterloo
Waterloo, ON
Canada

ISSN 2192-6255   ISSN 2192-6263   (electronic)
Studies in Applied Philosophy, Epistemology and Rational Ethics
ISBN 978-3-319-52514-3   ISBN 978-3-319-52515-0   (eBook)
DOI 10.1007/978-3-319-52515-0

Library of Congress Control Number: 2017933541

Printed on acid-free paper

This Springer imprint is published by Springer Nature
The registered company is Springer International Publishing AG
The registered company address is: Gewerbestrasse 11, 6330 Cham, Switzerland

Preface

Design and Society is the name of a course offered by the Centre for Society, Technology and Values at the University of Waterloo. The main aim of the course is to discuss what constitutes good design from social rather than technical perspectives. That is, the course involves an examination of ways to assess designs that are social instead of technical in nature.

One of the challenges in presenting this course has been the lack of materials that were both relevant and accessible to its students. Of course, the literature in this area is plentiful, at least, on certain topics. However, the course is intended for an audience of undergraduates from any faculty, who may well have no background in technology–society studies and whose only exposure to the field may well be this course. As a result, a course based on readings of seminal articles in the field is inappropriate.

In the end, the best course of action was to present a selection of concepts framed in a plain but pertinent fashion. By pertinent, I mean concepts that lend themselves to the practical business of design evaluation. Each module of the course is intended to equip students with the means to think clearly and critically about the social nature and impact of technological design. After several years of teaching, revising and adjusting, a coherent and significant set of concepts has been assembled.

At this point, a book in which these concepts are systematically presented and explained makes sense. In the absence of a set of readings, a book would provide students with a reference point useful for reinforcing the material. It would allow classes to incorporate less lecturing and more examination of nuances, cases, and implications. That book now lies in your hands or on your screen.

Because the book is aimed at an audience without any particular background in technology–society studies, it should also be suitable to anyone in the general public who is interested in these issues and looking for an accessible and general overview. As the world we live in becomes more and more an environment of our own design, social assessment of design becomes ever more relevant and important. General interest in this topic is high and will likely only increase. I hope that this book will prove useful and instructive to anyone for whom technology and society is a topic of present concern.

The book is organized into three parts, each part covering concepts representing a distinct perspective on technology–society relationships when it comes to good design. The first part provides an empirical perspective. That is, it examines how good design relates to knowledge of the social world. It begins with a presentation of two contrasting views of what sort of social knowledge is relevant to good design. On one view, represented by the noted industrial designer Dieter Rams, good design is a professional concern tied to a social mission, namely to make the world a more humane place.

On another view, represented by the noted decision theorist Herbert Simon, good design is primarily a matter of optimal problem solving, for which expertise in social studies is crucial. The remainder of the first part takes up Simon's perspective, examining how concepts from social psychology, anthropology and economics can assist in design assessment.

In the second part, the book returns to a perspective of the sort recommended by Rams. There, designs are assessed not only as means to whatever ends, but by the ends that they achieve or are intended to achieve. Concepts for this sort of assessment are drawn from applied ethics and from sociology, e.g., social contracts and social agendas.

In the third part, the book turns to judgments of designs where uncertainty about their ends is significant. That is, designs must often be assessed when important information about their potential impacts is unclear. The concepts of risk and fairness are adapted from applied ethics for this purpose.

Throughout this presentation, the task set out for the reader is just the same: To assess designs by applying the concepts at hand. To keep the presentation straightforward and practical, some (or many) matters of nuance are left out or only glanced at. Also, many additional concepts that might be selected and applied have been omitted. However regrettable in some respects, this simplification is necessary to keep the material at hand tractable and coherent.

As to the subject matter, the book takes a broad view of design but tends to focus on examples that are familiar and relevant to students in technically oriented fields. So, most examples of design concern functional objects or software services. Perhaps the only notable exception is provided in the chapter on social spaces in part two, where public structures and architecture are featured. However, the concepts explored in this book should be applicable more broadly as well.

I have also made an effort in this book to feature people as well as their things. It is easy, and appropriate, to fill a book on technological design with pictures and accounts of technology. However, technology remains an essentially human endeavor. As a result, I have tried to talk about and emphasize the people who are behind the ideas featured in this book. It is, after all, a book not about technology as such but about its relationship with people.

Thanks for the input and opportunities that resulted in this book go to Norman Ball, the former director of the Centre for Society, Technology and Values who recruited me to teach this course. Although I have changed some of the details of its curriculum, the course remains focused on the topic he set out for it, that is, good design from a social perspective.

Thanks also to Paul Thagard, my Ph.D. supervisor, whose guidance and advice has been indispensible to my scholarly development and academic career. Paul's facility with writing and researching has been a model for my own efforts. I hope that his example is reflected in this work.

Thanks to Lorenzo Magnani and Nancy Nersessian for supporting me in my scholarly advancement. The Model-Based Reasoning conferences organized by Prof. Magnani have provided an excellent opportunity for professional development and a challenging venue for the development of some of the ideas featured here.

Thanks also go to Scott Campbell, Director of the Centre, for his support and assistance with the Design and Society course. I am grateful to Karl Griffiths-Fulton, Graeme Epps, and Wendy Stocker for their work as teaching assistants for the course, and their feedback on its material and delivery.

I am also grateful to the many students who have taken this course over the last decade. Waterloo is fortunate to attract students of intelligence and energy, and their feedback on the selection and presentation of the materials in the course have been indispensible to the composition of this book.

Finally, I am indebted to my family, Julie and Corinna, for their love and forbearance during my time working at the Centre and, of course, in general.

Waterloo, Canada   Cameron Shelley
2016

Contents

What Is Good Design?

Good Design

What does the expression *good design* mean? What qualities distinguish good designs from poor ones? Think of some examples of good designs are consider what makes them good. Think of some bad designs and consider what makes them bad.

These questions usually draw out a variety of answers. A good design may be one that is *safe*, *efficient*, or *beautiful*. Or, it may be *useful*, *durable*, or *easy-to-use*. And, everyone has their own list of designs that they find good and others that they find awful and frustrating. Sometimes, a design can be good in one way and not good in another. Some designs will be considered good by some people and wretched by others. If you survey many people with these questions, you will realize that the matter of good design is far from simple, obvious, or uncontroversial.

In this situation, we may turn for help to the history of design. That is, we can examine designs that have proven popular, with the public or with critics. Also, we can canvas the opinions of successful designers, who often have very definite views on the subject. Let us begin, then, with a successful design, namely a garbage can called the *Garbino*.

Case Study: The Garbino

The Garbino is a household garbage can designed by Egyptian-Canadian designer Karim Rashid in 1997 (Fig. 1). It is produced by the New York design firm Umbra and has proven to be consistently popular in the North American market.[1]

The Garbino has a number of distinctive characteristics. Mr. Rashid wanted the can to be useful as a garbage receptacle, of course, but also pleasing to the eye. For

---

[1]For further information about the garbino, see Lidwell and Mancassa (2011) and HGTV Canada (2011).

© Springer International Publishing AG 2017
C. Shelley, *Design and Society: Social Issues in Technological Design*,
Studies in Applied Philosophy, Epistemology and Rational Ethics 36,
DOI 10.1007/978-3-319-52515-0_1

these reasons, he avoided the typical cylindrical shape in favor of a curved profile that is more "geometrically rich." To emphasize this point, the name *Garbino* refers to the actress Greta Garbo, who was known for her "sexy" curves.

In addition, the Garbino was designed to be easy to manufacture and inexpensive. It is made of polypropylene via injection molding, which allows it to be made in large numbers at a small cost per unit. It retails for only $8. In addition, the plastic composition is easy to tint, allowing it to be made in many colors to suit any taste, and can be given a shiny finish to help make it attractive and clean looking.

Besides doing well in the market place, the Garbino has been well received by critics. It has been added to the collection of the Museum of Modern Art in New York.

The Garbino has other qualities to consider besides its appearance and composition. It can also be stacked without locking together, unlike cylindrical cans, which makes stacked cans easy to separate. In addition, the bottom has a concave profile so that liquids do not build up in hard-to-reach bottom edges, thus making it easier to clean than conventional receptacles.

Q: In what ways is the Garbino a good or simple design? In what ways is it not?

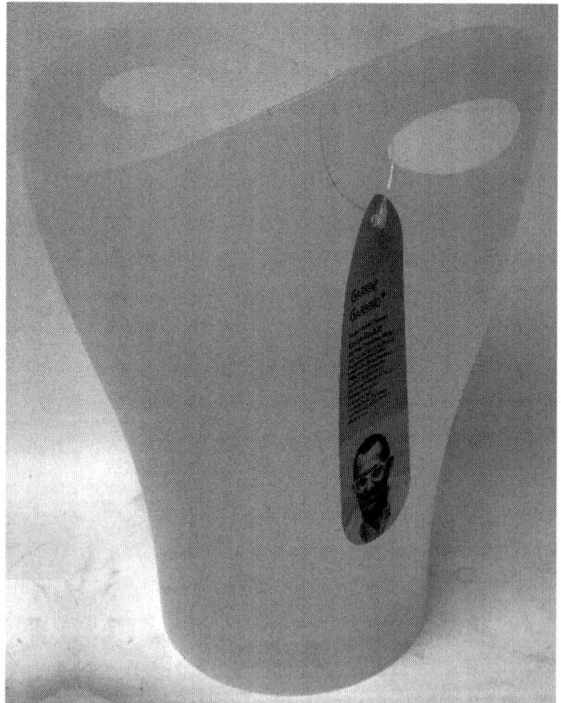

Fig. 1 The Garbino garbage can, by Umbra. Designed by Karim Rashid. Photo by Cameron Shelley

Dieter Rams

The Garbino provides an instructive example of how a contemporary and everyday design might be assessed. It is now appropriate to look at some general principles of good design. For this purpose, we can turn to Dieter Rams, a well-known and highly successful industrial designer of the 20th century (Fig. 2). In addition to being notable in his own era, the work and ideas of Dieter Rams continue to influence designers today.

Before discussing Rams's principles of good design, it is instructive to consider his record as a designer. Rams was born in Wiesbaden, Germany in 1932. His father was a radio engineer and his grandfather was a master joiner who made traditional furniture by hand. Rams said that he was greatly influenced by both men, most especially by his grandfather's commitment to craftsmanship in his work.[2]

In 1946, Rams studied at the Handwerker-und-Kunstgewerbeschule ("College of Manual and Fine Arts") in Wiesbaden where he trained as a carpenter and an architect. This mix of hands-on training and theoretical instruction suited Rams and was a hallmark of the College. The College emphasized *modernist* design as it had developed earlier in Germany and subsequently in the United States. He graduated with a degree in architecture in 1953.

In 1955, Rams joined the German electronics firm Braun. There he performed excellently and became head of product design in 1961. In his practice, he applied modernist design principles to home appliances of all sorts. He designed and co-designed more than 500 products during his career, from hairdryers, coffee-makers, sound systems, and book shelves to TV sets.

He has earned much recognition for his work. In 1968, he was given the "Royal Designer for Industry" award by London's Royal Society for the Arts. In 1996, he was given the "World Design Medal" by the Industrial Designers Society of America.

After his storied career, he retired from Braun in 1997.

Rams has written about the principles that guide his approach to design. Of particular significance are "Omit the unimportant"[3] and "The ten principles of good design" (often called the "Ten Commandments").

These writings are discussed below.

Minimalism: Less but Better

Rams's whole approach to good design is centered on minimalism. He encapsulates this approach in the Tenth Commandment as follows:

---

[2]Rams's education, work, and "Ten Commandments" are discussed in Lovell (2011).
[3]See Rams (1984).

Fig. 2 Dieter Rams, one of the most productive and influential modernist industrial designers, and author of the *Ten principles of good design*. Photo by Jonas Forth. URL: https://flic.kr/p/7yjupe

> Less but better—because it concentrates on the essential aspects, and the products are not burdened with inessentials.

Note the phrase "Less but better", which is a translation of the original German phrase "weniger aber besser." This expression suggests that good design is crucially a matter of leaving things out. A design is good only when features cannot be taken away from it without making it worse. Conversely, if a feature can be removed from a design without adversely affecting it, then the feature is inessential and should indeed be left out.

A straightforward illustration of this idea comes from Steve Jobs, one of Rams's biggest fans. Jobs led Apple's Macintosh design project in the early 1980s and mandated that it should use a graphical user interface, which is now the basis of all popular computer operating systems such as Apple macOS and Microsoft Windows (Fig. 3). As a part of this interface, users of the Macintosh would use a mouse to point at locations on the screen. Before that time, users would use cursor keys (left, right, up, down) to move a cursor around the screen. Jobs decided that users should have only one way to move the cursor, that is, by moving the mouse. To force users to work this way, Jobs insisted that the Macintosh keyboard would not have cursor keys at all. In that way, even users who were initially uncomfortable with the new interface would be required to use it.[4]

This design appears to follow Rams's dictum: Having only a mouse to use for pointing the cursor is less than having both a mouse plus a set of cursor keys. Since

---

[4]See Levy (1994).

Fig. 3 Original Macintosh 128 k, made in 1984. Note the absence of cursor keys on the keyboard. Photo by Marcin Wichary. URL: https://flic.kr/p/4jA1sX

the mouse is essential for use of the graphical interface, the cursor keys are thus inessential and should be left out.

Of course, cursor keys soon reappeared on Macintosh keyboards. In part, this was due to conservatism among users. They were accustomed to cursor keys and complained about their absence in the new design. In addition, cursor keys are more convenient for certain applications, such as filling in blanks in forms on the screen, a function that Steve Jobs probably did not consider when thinking about the Macintosh interface.

These points illustrate some forces that act against the minimalism that Rams advocates, namely backward compatibility with older designs and differences in perspective on what is essential or not to a design. Minimalism seems most appropriate for novel designs but much design work is actually re-design work where retention of old features must be considered.

Rams's appeal to the essence of a design seems to assume that there is some ideal form that captures what a design should be like. In that case, the job of designers is to understand this ideal and make a design that conforms to it. However, such ideals are notoriously elusive and open to disagreement and change. So, a general problem for Rams's approach to good design based on minimalism will be how the essence of a design is to be identified.[5]

Q: Think of some good designs. Do they satisfy Rams's principle of *less but better*? What designs could be improved by application of this principle?

---

[5]See Shelley (2015).

Omit the Unimportant

Rams tried to explain his considered view of good design in a number of ways. Following his minimalist views, his writings are always short and succinct. One of the best expositions is in his article "Omit the unimportant",[6] which is characteristically brief at three pages long. Here, we will consider some of the principles that Rams presents.

> One of the most significant design principles is to omit the unimportant in order to emphasize the important... [e.g.,] items that have unconstricted obvious-seeming functionalism in both the physical and the psychological sense. Therefore, products should be well designed and as neutral and open as possible, leaving room for the self-expression of those using them.

Here, by "important", Rams means important for users, and thus not for designers or others involved in production. Good designs emphasize features that are important to their intended audience. By implication, features that are not important to users should be de-emphasized or omitted altogether. By so doing, Rams explains, users are more likely to understand what a design is supposed to do and how that design may be used to achieve what they want.

In fact, this principle goes beyond simple minimalism and suggests that features of a design may be ranked in terms of importance to users. Features that are highly important are made prominent while features that are less important may be de-emphasized or even hidden away. Rams provides the Braun *Atelier* sound system as an example. In the caption, Rams remarks that (Fig. 4)

> With its blend of order, neutrality, and mobility, this Braun hi-fi construction is the expression of Braun design philosophy. The operating components that are seldom used are set in the back of the stereo. The entanglement of wires is concealed behind a cover so that the stereo may be placed in an open space.

Q: How does this system satisfy "omit the unimportant"? What other designs satisfy this criterion? Fail to satisfy it?

This principle also appears as the sixth of Rams's "Ten Commandments of Good Design":

> Good design is unobtrusive. Products fulfilling a purpose are like tools. They are neither decorative objects nor works of art. Their design should therefore be both neutral and restrained, to leave room for the user's self-expression.

In this formulation, Rams assumes that good design allows users to achieve their goals ("self-expression") as straightforwardly as possible. The problem he sees here is the intrusion into design of the goals of the designer. Here, he warns designers

---

[6]Rams (1984).

Fig. 4 Braun *Atelier* stereo
system. Photo by Nick Wade.
URL: https://flic.kr/p/7htVHA

not to think of themselves as artists whose goal is to express themselves in their
works. Only once designers have suppressed their own preferences can their work
concentrate on what is important, or not, to users.

Clarity

When important features have been identified and ranked—and unimportant ones
eliminated—there is still the problem of communicating the remaining features to
users in a proper way. For Rams, the only proper way is with clarity:

> ... items should be designed in such a way that their function and attributes are directly
> understood... Design riddles are impudent and products that are informative, understand-
> able, and clear are pleasant and agreeable.

Ideally, Rams says, a product should be self-explanatory, that is, it should be clear to users how it is properly used, without frustration or the need to refer to horrible user manuals.

Rams's illustration is the ET 66 calculator. (ET is short for "Elektronische Taschenrechner" or electronic pocket calculator) (Fig. 5). Here is what he says about it in the caption:

> Input functions of this Braun pocket calculator are coded in green, output functions are in red. The clear legible display and the arched keys facilitate fast computations. If the power switch is left on, it automatically turns itself off after six minutes.

You might say that this calculator is clear in the sense that it interface is "legible", that is, easy to read. However, like a message in any language, the user must know the language in order to understand it.

Fig. 5 Braun ET66 pocket calculator. Photo by Cameron Shelley

Designers sometimes say that a product or line of products rely on a particular *design language*, that is, a consistent way of using shapes, colors, textures, behaviors, etc., to communicate themselves to users.[7]

> Q: What are some elements of the design language of the ET 66? How do they help to make the design clear or legible?
> Q: What other designs are clear? Unclear?

This criterion is given as the fourth of Rams's "Ten Commandments of Good Design":

> Good Design makes a product understandable. It clarifies the product's structure. Better still, it can make the product talk. At best, it is self-explanatory.

Of course, getting products to "talk" by means of design is a demanding task. Creativity, experience, tenacity, ability, and diligence are necessary on the part of designers.

Restraint

In addition to promoting understanding through clarity, good designs should avoid making emotionally intense appeals to users.

> The latest design trends are intended to evoke emotions by trivial, superficial means ... The issue is stimuli: new, strong, exciting, and therefore aggressive signals. The primary aim is to be recognized as intensely as possible. The aggressiveness of design is expressed in the harshness of combat to attain first place in people's perception and awareness and to win the fight for a front place in store display windows.

Rams's point is that designs can sometimes succeed in the marketplace, at least for a while, by being merely showy or attention-grabbing. This practice is an error in at least two ways:

1. Pursuit of showy features takes effort and attention away from making designs clear and useful;
2. Pursuit of showy features is underhanded and exploitive, involving "the ruthless exploitation of people's weaknesses for visual and haptic signals...".

In this sense, restraint is about both the design and the designer. Regarding designs, restraint is an appeal for plain layout, neutrality of appearance, and the absence of decorative elements. Regarding designers, restraint is an appeal for focus

---

[7]See Sudjic (2008).

Fig. 6 Braun ABW 41 wall
clock. Photo by Phrontis.
URL: https://commons.
wikimedia.org/wiki/File:
Braun_ABW41_(schwarz).
jpg

on the needs of users and exclusion of the needs of designers, where those are at odds.

Rams also says that designers, who are supposed to be innovative, may confuse merely showy novelties with genuine innovations. He blames this problem on trends in the design industry—e.g., rivalry for professional recognition or attempts to gain a reputation for cleverness. We will return to the issue of innovation shortly.

The Braun wall clock illustrates the restraint recommended by Rams (Fig. 6). Here is what is says about it in the caption:

> This Braun wall clock exemplified aesthetic functionality and adaptibility with an economical use of its resources and a distinctly legible clock face, and has a metal casing and a plexiglass cover.

> Q: In what way is this design restrained? How does it compare to other wall clocks?
> Q: What other designs are restrained? Not restrained?

The term *restraint* does not correspond exactly to any one of Rams's "Ten Commandments of Good Design." However, it does capture Rams's view that designers owe it to society to prevent their work from becoming merely showy:

> Our culture is our home, especially the everyday culture expressed in items for whose forms I am responsible. It would be a great help if we could feel more at home in this everyday culture, if alienation, confusion and sensory overload would lessen.

Rams thinks that good design is to be assessed not only in technical or user-centered terms but in terms of society as well.

Progressiveness

Rams argues that designers have a social responsibility and, therefore, a pro-social mission:

> Design means being steadfast and progressive rather than escaping and giving up. In a historical phase in which the outer world has become less natural and increasingly artificial and commercial, the value of design increases. The work of designers can contribute more concretely and effectively toward a more humane existence in the future.

Here, Rams points out that world in which we live becomes more and more a world of our own design rather than a natural world. Today, most people in the world live in cities or towns rather than in the wilderness. This trend will continue into the foreseeable future. Rams argues that, as we live in a more artificial world and are ever more dependent on technology, the responsibility of designers to make that world a humane one increases also.

How designers can achieve these progressive aims Rams does not explore in detail. However, he does mention his own aims:

> I work in the hope of designing objects that are useful and convincing enough to be accepted and lived with for a long time in a very obvious, natural way. But such objects do not fit into a world of vandalism, aggression, and cynicism. In this kind of world, there is not room for design or culture of any type.

By *progressive*, then, Rams seems to mean the following:

1. Designs are progressive if they fit into and promote social stability;
2. Designs are not progressive if they promote social conflict.

Case Study: The Vitsoe 606

Rams does not provide an explicit illustration of progressive design at Braun, but we might consider his "606 Universal Shelving System" designed for Danish furniture-maker Vitsoe (Fig. 7).

It could be said to be progressive in the sense that it provides for easy and peaceful participation in a harmonious work environment. The shelving allows users to adapt their space for use as an office, office work being a promising and safe, middle-class occupation.

Contrast Rams's bookcase design with the Carlton bookcase (Fig. 8) designed in 1981 by postmodernist designer Ettore Sottsass (Fig. 9). Rams would doubtless consider it a design abomination as a bookcase. It is stridently colored, relentlessly

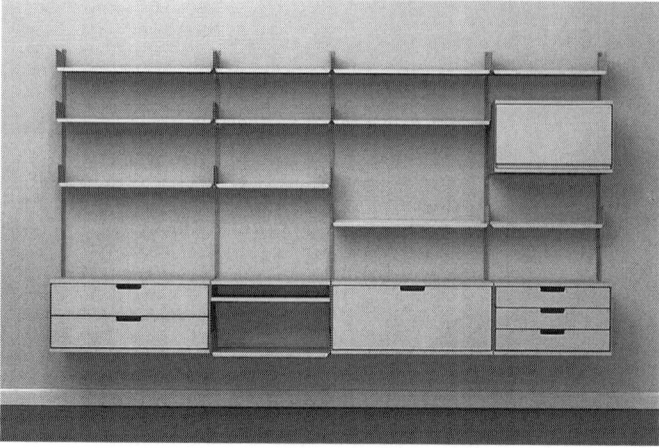

Fig. 7 Vitsoe 606 Universal shelving unit. Photo by Rams-ethos. URL: https://commons.wikimedia.org/wiki/File:Vitsoe606.jpg

Fig. 8 *Carlton* bookcase by Ettore Sottsass, 1981. Photo by Memphis Milano. Detail of URL: https://flic.kr/p/nydZue

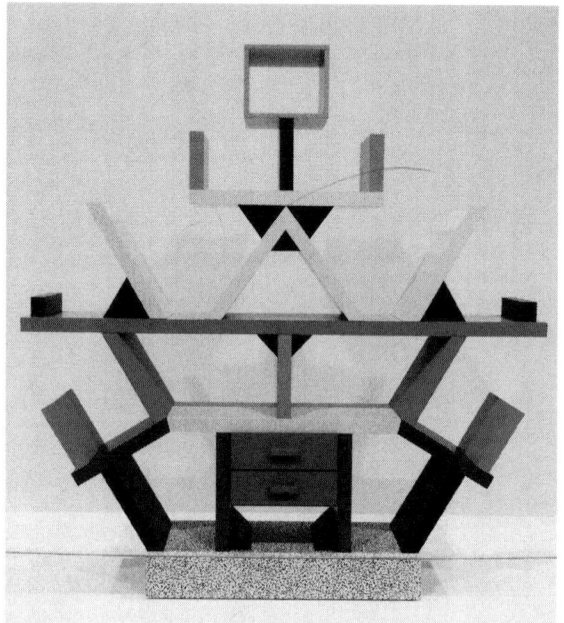

Fig. 9 Ettore Sottsass (1917–2007) in 1969. Sottsass was a prominent postmodernist designer perhaps best known for his work on Olivetti typewriters. Photo by Guiseppe Pino. URL: https:// upload.wikimedia.org/ wikipedia/commons/1/1c/ Ettore_Sottsass_1969.jpg

jagged, and unlikely to harmonize with most library settings. As a bookcase, it is likely to cause frustration and even argument. Rams would consider it, at best, a misplaced artwork and, at worst, an act of vandalism by the designer.

Q: Is the Vitsoe 606 progressive? Is the Carlton not progressive?
Q: What other designs are progressive? Not progressive?

Innovativeness

Although this quality is not mentioned explicitly in *Omit the unimportant*, Rams also includes innovation among the qualities of good design. The first of his "Ten Commandments of Good Design" is:

Good design is innovative. The possibilities for innovation are not, by any means, exhausted. Technological development is always offering new opportunities for innovative design. But innovative design always develops in tandem with innovative technology, and can never be an end in itself.

Technology is constantly bringing forward new ways of doing things, which designers should be ready to employ in their work.

It is almost a relief to hear that good design is not merely about minimalizing things or being restrained. It can also mean adding new things to designs.

However, Rams cautions designers not to confuse innovations with the development of mere novelties that do not bring with them increases in functionality. Both innovative designs and novelties will be different than previous designs. The distinguishing feature is that innovations make designs genuinely more useful. Although this distinction sounds simple enough in principle, it can be difficult and contentious in practice.

Case Study: The Bruno

Designers Jim Howard and Lori Montag have designed what they consider a better garbage can, the Bruno. The Bruno garbage can has the same basic function as any other trash can in that it is a container in which a trash bag can be placed for temporary storage of garbage (Fig. 10). In addition, the Bruno has an integrated vacuum cleaner with an input through a slot on the bottom, front edge. The idea is that the Bruno can use the integrated vacuum cleaner to pick up a pile of sweepings made by a broom, which would normally be handled by a dustpan. Thus, the Bruno would not require stooping like a conventional dustpan, nor close proximity to dirt. Also, the vacuum would not leave a small ridge of dust that dustpans often do.

Fig. 10 Bruno, the smart trashcan. Photo by Design Milk. URL: https://www. flickr.com/photos/designmilk/ 17723364979/in/photolist-d6ZFuY-bNqwXr-t19SWg-6mwCoT-6bLK9Y

There are some further functions. For example, the Bruno has a sensor that opens the lid when an object approaches so that its user does not have to handle a dirty lid. Also, it has a space for storage of extra bags and can notify owners by smartphone when the current bag is getting full. It has a battery and needs to be charged about once a month. The initial cost to Kickstarter backers was $159.[8]

Q: Is the Bruno innovative or merely novel, on Rams's view? Why or why not?

Q: What other designs are innovative? Not innovative?

Critique: Is Good Design Too Dull?

Rams's designs and principles have been widely influential in industrial design. For example, Steve Jobs of Apple was an admirer, as were other Apple designers such as Harmut Esslinger and Jonathan Ive. However, we have seen that Rams's principles are not self-evidently true and also not above criticism. Here, we will consider some critiques of Rams's views.

First, Rams's principles are largely negative. That is, they focus largely on the importance of subtracting inessential features from design, and why this is a good idea. At the same time, Rams's principles say little about how to add features to designs. He does point out (Commandment 1) that good design is innovative and responds to advances in technology. However, he offers little help in distinguishing innovations from novelties, other than to warn against mere enthusiasm for new things.

As a result, his advice seems a little one-sided. Or, perhaps there is simply no principled way to bring technological innovations into design work!

Second, heavily simplified designs can be considered dull. The architect Robert Venturi, speaking about the mania for minimalism in modernist architecture, said that, "less is a bore".[9] Rams's preference for shades of white, grey, and black could be considered rather bloodless. Designer Gadi Amit observes this from an exhibition of Rams's works[10]:

> Stepping back from the displays, the exhibition struck me as visually cold. Looking at Rams's work, one cannot ignore the color tonality, or to be more precise, the lack thereof. The mechanical precision, strict design language, and above all the gray mood are overwhelming. The work is timeless, created as if postmodernism had never arrived. With only a few colors other than gray, anemic wood selections, and the use of stark metal mesh, the

[8]Weiner (2015).
[9]Venturi (1962, p. 17).
[10]Amit (2012).

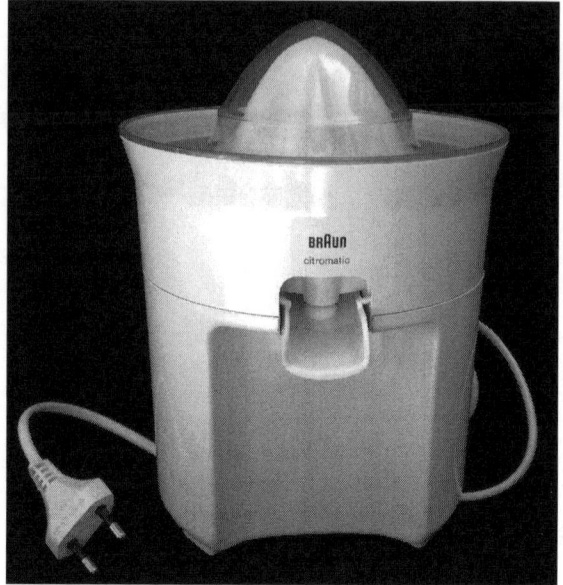

objects cast a subdued and lifeless shadow. These are efficient objects, beautiful in their purposefulness, yet they fail to enlighten or revive their surroundings.

Perhaps Amit was looking at a design like the Braun Citromatic (Fig. 11). On the face of it, the Citromatic is simply an off-white cylinder with a handle on one side and a point on the top. Although it may work well, its appearance is rather dull. Yet, it is the design that got Jonathan Ive of Apple excited about a career in industrial design![11] Perhaps Ive is just a special person.

Indeed, simple and functional designs can be quite deadening. Consider the fate of the Action Office designed in the 1960s by Robert Probst.[12] Probst designed a set of office furniture meant to be highly neutral, mobile, and reconfigurable. He thought this would turn an office from a rigid grid of walls and desks into a flexible, dynamic environment. Early versions were too elaborate and expensive and did not sell well, so the design was simplified into a set of rectangular walls and shelves made of cheap materials now known as the cubicle. These were knocked off by competitors and became very popular. By the 1980s, millions of office workers beavered away in large spaces, each one surrounded by their own dividers and cut off from their neighbors. Despite—or because of—being simple and functional, the cubicle has become a foundation of deadening and oppressive workplace design (Fig. 12).

---

[11]Lovell (2011).

[12]On the history of the cubicle, see Saval (2014).

Fig. 12 A lot of cubicles that seem to go on forever. Photo by Larsinio. URL: https://commons. wikimedia.org/wiki/Category:Office_cubicles#/media/File:Cubicle_land.jpg

Furthermore, this problem seems to threaten to undermine Rams's program for good design and society. It seems that other designers could take a spare and minimalist Braun design, jazz it up and add some extra (though superfluous) functions and colors, and thus out-perform the Braun design in the marketplace. Rams admits that people are susceptible to manipulation by designers in this way. It seems, then, that Rams's program for good design is futile, that is, it cannot achieve the broader social goals that he aims to achieve.

For example, consider Coda Automotive's little electric car (Fig. 13). It was an innovative electric car and among the first such cars to be put on the consumer market. Its appearance is quite unremarkable—that is, a bore—compared with later entries such as the Nissan Leaf or the Tesla. In 2013, it was quietly withdrawn from the market altogether after selling only 117 units.[13]

Good design is important. As Dieter Rams pointed out, the world is more and more an artifact of our own creation. Its designers bear a substantial responsibility to ensure that it is a good world and that it improves over time. Yet, the meaning of *good design* is neither straightforward nor self-evident. Rams's minimalist approach provides us with some insights and some serious themes to consider. However, it seems neither completely clear nor comprehensive.

What is clear is that good design is not solely a matter of technical prowess. It is also a matter of attention to the broader social context in which designs are received and adapted. The nature of this social context is explored further in the following chapters.

---

[13]Bullis (2013).

Fig. 13 CODA electric car exhibited at the 2012 Washington auto show. Photo by Mario Roberto Durán Ortiz. URL: https://commons.wikimedia.org/wiki/File:CODA_sedan_WAS_2012_0835. JPG#/media/File:CODA_sedan_WAS_2012_0835.JPG

References

Amit, G. (2012, Jan 4). *Gadi Amit offers 4 lessons from Dieter Rams's successes, and failures.* Retrieved January 18, 2012, from FastCompany: http://www.fastcodesign.com/1665747/gadi-amit-offers-4-lessons-from-dieter-ramss-successes-and-failures

Bullis, K. (2013, May 1). *Maker of world's most boring cars stops making cars.* Retrieved September 18, 2015, from Technology Review: https://www.technologyreview.com/s/514501/maker-of-worlds-most-boring-car-stops-making-cars/

HGTV Canada. (2011, January). Garbino. *Design DNA.* Toronto, ON, Canada.

Levy, S. (1994). *Insanely great: The life and times of Macintosh, the computer that changed everything.* New York: Viking.

Lidwell, W., & Mancassa, L. (2011). Garbino trash can. In *Deconstructing product design.* Beverley, MA: Rockport Publishers.

Lovell, S. (2011). *Dieter Rams: As little design as possible.* London: Phaïdon.

Rams, D. (1984). Omit the unimportant. In V. Margolin (Ed.), *Design discourse: history, theory, criticism.* Chicago: University of Chicago Press.

Saval, N. (2014). *Cubed: A secret history of the workplace.* New York: Doubleday.

Shelley, C. (2015). The nature of simplicity in Apple design. *The Design Journal, 18*(3), 439–456.

Sudjic, D. (2008). *The language of things.* London: Penguin.

Venturi, R. (1962). *Complexity and contradiction in architecture.* New York: Museum of Modern Art.

Weiner, S. (2015, June 5). *Say goodbye to dustpans: The vacuum trashcan is here.* Retrieved September 9, 2015, from FastCompany: http://www.fastcodesign.com/3045901/say-goodbye-to-dustpans-the-vacuum-trashcan-is-herechapters.

Rational Design

Abstract Instead of viewing it as a professional pursuit with a central, social mission, good design may also be viewed as a technical exercise in optimal problem solving. This understanding of good design is the dominant perspective in most engineering and other technical programs. Decision theorist Herbert Simon's influential characterization of good design as optimal problem solving is described and explored. On this view, good design is *rational design* in the sense that any rational being would want designs that are objectively optimal. Simon's view is highly general in the kind of design problems it applies to and what constitutes an optimal solution. However, it also involves unrealistic assumptions about the knowledge and objectivity of problem solvers. A more realistic characterization of problem solving is provided with the concept of *boundedly rational design*. On this view, increasing knowledge is key to good design. This view sets the scene for the following chapters in which knowledge derived from the social sciences is examined for its relevance to good design.

The Best Solution

In the previous chapter, we reviewed Dieter Rams's brief account of good design. From that account, it seems clear that Rams conceived of design primarily as a professional role. As a result, his remarks are framed as advice to professional designers. For example, they concern good professional practice, such as putting the interests of users ahead of designers, and the social mission of designers, such as making the world more humane.

Although this approach is perfectly legitimate, it has its limitations. For example, Rams assumes that his audience will have some experience with design and particular intuitions about it. Thus, he refers to concepts, such as the *essence* of a design, without ever defining them. On the whole, his program of design assessment derives very much from his personal history. This fact is hardly regrettable, as Rams is a noted and successful designer. However, it may leave readers with questions about the objectivity or generality of his recommendations.

© Springer International Publishing AG 2017
C. Shelley, *Design and Society: Social Issues in Technological Design*,
Studies in Applied Philosophy, Epistemology and Rational Ethics 36,
DOI 10.1007/978-3-319-52515-0_2

A very different perspective on design assessment is offered by Herbert Simon. Simon had an interest in design but as a theoretician rather than a practitioner. He attempted to characterize design and design assessment from a highly general and objective standpoint. On his view, design is simply a kind of problem solving, that is, changing the world so as to overcome some problem. Good design, put briefly, is solving a problem *in the best possible way*.

As such, design is something that anyone might do at any time and not primarily a professional activity with a particular social mission attached. This view of good design, with its general and impersonal perspective, has been highly influential, especially in technical disciplines.

Yet, as you might expect, it also faces a number of serious challenges. For example, strictly speaking, his account of good design would require a nearly God-like omniscience and powers of foresight. Also, a completely objective description of good design may be hard for actual designers to achieve, given that they all have their own backgrounds and perspectives. Simon was aware of these difficulties and attempted to deal with them, as we will see.

In this chapter, we will examine Simon's general theory of design and his depiction of good design as *rational* design. Then, we will explore how this perspective applies in practice. It will be helpful to begin with a case study.

Case Study: The Open Office

In the previous chapter, the cubicle was presented as an example of office space design. In some respects, the cubicle is a reasonable solution to the problem of providing work space for "knowledge workers", that is, people whose work mainly involves applying and processing information. The cubicle provides privacy so that knowledge workers can concentrate on their assigned tasks, often involving a computer, while minimizing distractions from their surroundings. Cubicles are also flexible in form and restrained in presentation, allowing for personalization and adaptation to different tasks. In addition, cubicles are cheap and depreciation in their value over time can be written off as a business expense.

However, in the last decade or two, many employers have realized that cubicles pose substantial problems for knowledge workers. In particular, isolation imposed by cubicles may actually hamper productivity by preventing knowledge workers from communicating well with each other.

In order to overcome this defect with cubicles, some offices have moved to the *open office* concept, in which interpersonal walls are eliminated so that workers can easily interact (Fig. 1). These interactions, whether planned or casual, are thought to be crucial in helping knowledge workers to getting their work done.[1]

---

[1]Saval (2014).

Fig. 1 An open office at Lyons Architects. Photo by Peter Bennetts. URL: https://commons.wikimedia.org/wiki/File:Lyons_Architects_Office.jpg

Furthermore, the open office makes the activities of workers easier to police.[2] Workers whose whereabouts and activities are easily visible can be tracked more effectively by managers. This transparency helps mangers to ensure that workers are not simply wasting time.

In addition, employees in open offices can police one another through peer pressure, perhaps frowning at or otherwise making clear their disapproval of colleagues who are not obviously pulling their weight in the office.

On this account, adoption of the open office design seems rational. Knowledge workers often need to exchange information in order to do their work. The open office facilitates such interactions and, in that respect, seems clearly better than the cubicle. Add the greater ease of policing work in the open office and the design seems easily to outperform the alternative.

However, adopters of the open office design did not anticipate some of its consequences for employees. For example, increasing the ease and frequency of employee interactions does not ensure increased productivity. Often, interactions are distracting. Consider an observation by Lindsey Kaufman relating what happened to her on her first day in the new open office design with her employer, an ad agency[3]:

---

[2]The Economist (2016).
[3]Kaufman (2014).

> Our new, modern Tribeca office was beautifully airy, and yet remarkably oppressive. Nothing was private. On the first day, I took my seat at the table assigned to our creative department, next to a nice woman who I suspect was an air horn in a former life. All day, there was constant shuffling, yelling, and laughing, along with loud music piped through a PA system.

She said that she purchased noise-cancelling headphones for work the same day.

In general, employees working in open-plan offices often complain about difficulties in dealing with distractions, and a lack of sound and visual privacy.[4]

Kaufman also observes that even agreeable interactions are no guarantee of increased productivity:

> The New Yorker, in a review of research on this nouveau workplace design, determined that the benefits in building camaraderie simply mask the negative effects on work performance. While employees feel like they're part of a laid-back, innovative enterprise, the environment ultimately damages workers' attention spans, productivity, creative thinking, and satisfaction.

As to the second assumption, being constantly under the scrutiny of others can become stressful and counterproductive, as Ms. Kaufman also notes:

> As an excessive water drinker, I feared my co-workers were tallying my frequent bathroom trips.

In other words, employees may find the social policing of open offices to be a source of stress and thus a brake on productivity.

In the end, the open office design is problematic. Although it does have qualities that would tend to enhance the productivity of knowledge workers, it also has qualities that detract from productivity in ways that were not anticipated by the designers.

> Q: Was the open office design a rational response to the problems of cubicles?
> Q: In what ways does the design of your workplace enhance or hamper your productivity?

We can make the concept of rational design clearer by examining Simon's account of it.

Herbert Simon

Herbert Simon (1916–2001) was born in Milwaukee, Wisconsin and graduated with a Ph.D. in Political Science from the University of Chicago in 1943 (Fig. 2). After serving as at the Illinois Institute of Technology, Simon took up a position at the

---

[4]Varjo et al. (2015).

Carnegie Mellon University (then Carnegie Tech) in 1949, where he remained until his death in 2001.

Simon's central interest was in human reasoning, especially decision-making. His early work concentrated on decision making in organizations, where he worked on models of decision-making that were psychologically plausible, that is, that took into account idiosyncrasies and limitations of human psychology. As such, Simon undertook psychological studies of human reasoning and modeling of it through computer simulations.

Simon viewed design as important because of its centrality to decision-making. According to Simon, decision-making involves gathering information about a problem, examining alternative solutions to it—designs—and then determining which solution is best.

He was also interested in why people sometimes fail to arrive at the best solutions.

Simon received numerous awards for his work, including the American Psychological Association Award for distinguished contributions to psychology (1969), the Association for Computing Machinery's Turing Award for contributions to artificial intelligence (1975), and the Nobel Memorial Prize in Economics for his work on organizational decision-making (1978).

Simon summarized the main ideas of his work in *The Sciences of the Artificial* (1981), which are presented in the following sections.

Fig. 2 Herbert Simon (1916–2001) studied and.built computer models of human reasoning. He received the 1978 Nobel Memorial Prize in Economics for his research into organization decision-making. Photo courtesy of Bettman/Getty Images

Design Problems

Simon begins by developing a broad definition of design. In his view, design is about the invention of artifacts, where *artifact* is understood in a very general way, to mean anything that does not occur naturally. To understand design better, it helps to begin by distinguishing artifacts from natural objects and how academic disciplines are divided depending upon which sort of thing they involve. Simon characterizes this distinction in terms of essential differences between science and engineering.

1. *Science* tends to involve the study of natural things. Thus, one of the main objectives of science is to explain how natural things work.
2. *Engineering* tends to involve the study of artificial things, i.e., artifacts. Artifacts are, by their nature, not usually things that simply occur around us. Instead, they have to be made. Thus, engineering tends to concern how artifacts should be made and designed.

In other words, scientists study things that already exist, such as planets, gases, and animals. Biologists, for example might study frogs, which are things that occur in nature. By contrast, engineers study things that do not already exist but must be made, such as bridges, computers, and solar panels.

As such, scientists and engineers are faced with very different intellectual tasks. The main job of scientists is to explain how natural things work, e.g., how stars produce light and how frogs produce tadpoles. By contrast, the main job of engineers is to design artifacts that solve problems, e.g., a new way of converting sunshine into electricity.

Yet, Simon argues, many professions besides those labeled as "engineering" are centrally concerned with design of artifacts[5]:

> Everyone designs who devises courses of action aimed at changing existing situations into preferred ones. ... Schools of engineering, as well as schools of architecture, business, education, law, and medicine, are all centrally concerned with the process of design.

Here, Simon takes a very broad view of a design problem: A design problem is any situation that is to be changed to conform to the designer's preferences. In brief, design involves how we turn the world from the way it is into the way we want it to be.

In engineering, this view seems appropriate enough. A civil engineer, for example, might have the task of taking a given situation, like an uncrossable river, and turning it into a preferred situation, like a river that is crossable. This task might be accomplished, in part, by designing a bridge that spans the river and that people can cross.

However, people in other professions face similar problems.

---

[5]Simon (1981), p. 129.

Q: How is a doctor a designer, on this view? A novelist?

Doctors are best known for designing therapies. A therapy is a solution to the problem of changing a sick patient—the given situation—into a healthy patient—the preferred situation.

That Simon's definition of design applies equally well to engineering, medicine, novel-writing, and many other activities, speaks to its great generality.

The Design Environment

Having defined a design problem in a highly general way, Simon proceeds to define design solution also in a highly general way.[6] First of all, the *design environment* is divided into two components. The first component he calls the *inner environment*, which is the set of all alternative solutions that might be applied to a given problem.

The second component is the *outer environment*, which is the context in which all alternative solutions have to operate.

For example, if the design problem were to design a car, then the inner environment would be the set of all possible cars that could be made by the means available. Obviously, that would be an enormous set of alternatives!

The outer environment of this problem would be the roadways in which the car would have to operate. A car has to be able to traverse a number of different kinds of roads, in varying conditions, and under different sets of regulations.

On occasion, design problems can be solved quite directly. For example, the problem of designing a meal might be solved readily by simply reheating a microwave pizza.

However, many important and interesting design problems are not so easily solved. To design a healthy meal, for example, it is usually necessary to prepare some ingredients and combine them in a certain way, e.g., by following a recipe.

More generally, to solve such a problem, designers must figure out how to take a number of items—none of which can be used to solve the problem on their own—and identify some combination of them that does solve the problem.

For example, a heap of noodles, salt, oil, and cheese is not a meal as such. To make a dish, ingredients must be processed and combined in an appetizing way.

Difficulties that arise in configuring the inner environment to solve a problem is why design often so challenging.

---

[6]Simon (1981), pp. 134–135.

Rational Design

For an artifact to solve a given problem, it must successfully relate the inner and outer environments. In other words, it has to work in a way that changes the world from the way it is to the way that is preferred.

In order to speak clearly about how an artifact can accomplish its goal, Simon introduces some special terms. In plain English, these are the *means, laws,* and *ends*. He defines these terms as follows:

1. The *means* (or "*command variables*") identify the basic components that are available in given quantities for use in the problem situation. The means define the inner environment.
2. The *laws* (or "*fixed parameters*") identify the fundamental and unalterable facts that apply in the problem situation. These fixed parameters are something like the laws of nature: They do not change and everything in the situation obeys them.
3. The *ends* identify the parameters that the design has to satisfy in order to be considered acceptable. The ends describe people's subjective preferences and come in two varieties:

   a. The *constraints* typically identify thresholds that the design must exceed and tolerances that the design must never exceed.
   b. The *utility function* identifies how the goodness of the possible designs is to be measured or, at least, how we should decide on which of the competing designs to prefer. Usually, the utility function is regarded as a kind of optimization function. Thus, good design is presented as design that is optimal in some sense.

Using these terms, we can define a problem solution as the means that brings about the desired ends, and complies with the laws that apply to the situation.

In fact, we can go a step further and say that the *best solution* to a problem is the optimal solution, as defined by the utility function. The utility function is a procedure that takes the description of a problem solution and tells us how good a solution it is. The optimal solution is, by definition, the solution that gets the highest score according to the utility function.

This point is where the concept of *rationality* enters in. The optimal solution to a problem is also the rational solution, in the sense that any rational being would prefer an optimal solution to a sub-optimal one. That is just what it means to be rational.

Some examples will help to illustrate this rather abstract description.

Table 1 The diet problem described in general terms of Herbert Simon's theory of rational design

Types	Terms	Example: diet problem
Means	Command variables	List of foods and quantities
Laws	Fixed parameters	Prices of foods
		Nutritional contents
Ends	Constraints	Nutritional requirements
	Utility function	Cost of diet

Case Study: The Diet Problem

Simon illustrates this way of parsing the design environment by looking at the example of planning a diet.[7] Like an engineer or a doctor, a dietician is a designer whose task is to design diets, that is, a regime of food intake in order to keep someone healthy and happy. In this sense, a diet is the solution to a dietary problem. Using terms described above, the diet problem could be broken down as follows:

1. The means consist of whatever foods are available and in what quantity. The contents of a grocery store would be a typical set of dietary means.
2. The laws consist of the prices of the foods and the nutritional value of each kind of food, e.g., calories, vitamins, minerals, etc. These parameters are laws because they are facts about the world that are beyond the control of the dietician. That is, dieticians do not determine what foods are made of nor how much they cost.
3. The ends consist of:

   a. The nutritional requirements, e.g., a tolerance of no more than 2000 calories/day, a threshold of at least 10 mg of vitamin C, a prohibition against eating spinach perhaps, etc.
   b. The utility function might be a simple matter of lowest cost. In other words, the best diet is the one that uses solves the problem at the lowest price.

This situation is summarized in Table 1.

> Q: What are the means, laws, and ends that apply to the following designs? A clock? An essay? A game?

For the essay example, be careful not to confuse an essay with one of its physical manifestations. An essay is an abstract thing and not a particular physical object. After all, the same essay may be printed out on paper, displayed on a computer screen, or just stored in a computer file. The essay exists regardless of what physical form it may take. Thus, none of those things are means of having an essay, though they may be useful for composing one.

---

[7]Simon (1981), pp. 135–136.

Obstacles to Rationality

Simon's characterization of rational design as an optimal solution to a problem seems compelling. However, as noted above, it takes a rather God-like perspective on the world. After all, a design counts as rational if there cannot possibly be a better one. However, our knowledge of the world is limited and uncertain, so it is not often in practice that designers can say that a better solution than theirs is simply not possible.

Simon spells out some important obstacles to the achievement of rational design in practice. First, our knowledge of the inner and outer environments is not perfect. We may have only a partial or qualitative conception of how a given artifact or component will behave. Thus, it would be difficult to prove anything about its performance for certain.

For example, Simon notes that urban design often depends on statistical models of traffic.[8] When planners design highways, they rely on simulations of road usage that are gained from statistical studies of driving behavior. Such models are useful in the sense that they make predictions that are accurate enough for practical purposes, most of the time. However, because they are only approximations of the systems they represent, they will lead to incorrect conclusions on some occasions. A road system simulation, for example, may not take much account of bicycles or pedestrians.

Second, even if we had perfect knowledge of the parts of our artifacts, we may lack the computing capacity to forsee how different configurations will behave or will evaluate according to our utility criteria.

For example, Simon draws attention to the classic travelling salesman problem.[9] Consider a travelling salesman who must visit a list of cities. He considers the best route to be the one that involves the least mileage. It turns out that as the list of cities gets very long, it would simply take too long to compute the shortest route. This problem looms large for many package delivery services. Instead of finding the optimal solution, methods are used that will provide a route that its good enough for a driver to complete within a reasonable amount of time.

So, although the problem is well-understood and an optimal solution exists, it is just too hard to find it out in some cases.

Third, designers may not agree about either the nature of the problem at hand or the methods to be used to solve it. On the latter issue, Simon invites us to consider buildings designed by architects with very different methodological approaches.[10]

> An architect who designs buildings from the outside in will arrive at quite different buildings from one who designs from the inside out, even though both of them might agree on the characteristics that a satisfactory building should possess.

---

[8]Simon (1981), p. 37.

[9]Simon (1981), p. 139.

[10]Simon (1981), p. 150.

Assuming both methods are equally well justified, then it may be impossible to say which solution is better when those solutions are quite different in character.

On the nature of problems themselves, Simon's theory of rational design assumes that there is a unique and correct description of any given problem situation. However, designers may disagree on this point. For example, consider the fact that Inuit languages contain many words for different types of snow. More than that, these languages contain even more words for sea ice[11]:

> A lexicon of sea ice terminology in Nunavik (Appendix A of the collective work Siku: Knowing Our Ice, 2010) includes no fewer than 93 different words. These include general appellations such as *siku*, but also terms as specialized as *qautsaulittuq*, ice that breaks after its strength has been tested with a harpoon; *kiviniq*, a depression in shore ice caused by the weight of the water that passed over and accumulated on its surface during the tide; and *iniruvik*, ice that cracked because of tide changes and that the cold weather refroze.

This observation raises the question: How many kinds of sea ice are there? The best answer seems to be: It depends, on where you are and what you are doing. Note that it also depends on what tools you have, as in the kind of ice that gets its name from how it responds to being tested with a harpoon.

In reality, our conceptions of what our problems are and what should count as solutions depends on our own personal and cultural histories. It also depends on the means that are available. (Recall the old expression that if all you have is a hammer, then everything looks like a nail.) In such cases, there is no impersonal and objective fact of the matter. Where there is no agreement about the nature of a problem, it may not be possible to say that one solution is the best possible one.

Bounded Rationality

Because of these realities, we simply lack the God-like knowledge and power necessary to determine when a design is rational under all circumstances. Simon acknowledges these issues but argues that there is still a useful sense of rationality that can be applied to design assessment. A design may be considered rational within the boundaries of our knowledge and computing power. In other words, a design may be considered *boundedly rational* when it is as good as we know how to make it.

Claims that a design is boundedly rational are often justified in the following ways:

1. No other design is known to be more optimal;
2. The design rests on assumptions that have worked well before ("have stood the test of time");
3. The design rests on assumptions that are widely accepted, e.g., best practices or industry standards.

---

[11]The Canadian Encyclopedia (2015).

Such designs are boundedly rational in the sense that most reasonable people, in the same circumstances as the designer, would agree that the design is as good as we know how to make it right now.

Unintended Consequences

Simon's theory of bounded rationality does seem to provide an account of how designers in many disciplines approach their work. Further evidence in favor of his account can be found in cases of *unintended consequences*. Such consequences arise from designs that were rationally configured as far as the designers were aware, but that went wrong anyway due to a lack of knowledge or resources.

For example, in 2013, Toyota recalled 870,000 vehicles, including Camrys, Veznas, and Avalons due to bugs causing rogue airbag warnings and deployments. In fact, spiders had been building webs inside the air conditioning dispensers (Fig. 3). Water then condensed on these webs and dripped onto the air bag controllers. In turn,

Fig. 3 The yellow sac spider, better known to automobiles than to their designers. Photo courtesy of Richard Bartz/Wikimedia commons. URL: https://en.wikipedia. org/wiki/Cheiracanthium#/ media/File:Cheiracanthium_ mildei_male.jpg

that water shorted out the airbag controllers, causing warning lights to light up on dashboards in error. It could even cause driver's side airbags to deploy unexpectedly, something that occurs with explosive force.[12]

Clearly, Toyota engineers did not desire this to happen. However, not being entomologists, they did not know that spiders might find these vents to be attractive places for building webs. (However, Mazda did had similar issues beforehand with yellow sac spiders.[13])

This example shows that limitations of expertise can sometimes compromise the goodness of a design. The designers may have been perfectly competent by industry standards. The problem is that even competent designers have limitations that can sometimes lead to solutions that are not optimal, even though they appear to be to the best of their knowledge.

> Q: Can you think of other instances of unintended consequences that illustrate bounded rationality of design?

The view that good design is rational design has been highly influential, particularly in technological design disciplines. Although rational design is unattainable in practice, designers who adhere to this standard try to approximate it as well as they can. That typically means acquiring more and more knowledge. The more knowledge designers have, the more like an ideal, omniscient designer they become. Thus, the more rational are their designs.

From a technology-society perspective, following this lead means increasing our store of knowledge about how good design relates to social concerns. The following chapters illustrate how knowledge from the social sciences may be helpful in the assessment of good design.

References

Kaufman, L. (2014). Google got it wrong. The open-office trend is destroying the workplace. *Washington Post.* Dec 30. https://www.washingtonpost.com/posteverything/wp/2014/12/30/google-got-it-wrong-the-open-office-trend-is-destroying-the-workplace/ (accessed Jan 12, 2015).

Pearson, G. (2013). Happy Arachtober! There are spiders in your car!. *Wired.* Oct 18. http://www.wired.com/2013/10/happy-arachtober-there-are-spiders-in-your-car/ (accessed Aug 12, 2016).

Saval, N. (2014). *Cubed: a secret history of the workplace.* New York: Doubleday.

Simon, H. A. (1981). *Sciences of the artificial* (2nd ed.). Cambridge, MA: The MIT Press.

The Canadian Encyclopedia. (2015). Inuktitut words for snow and ice. *Historica Canada.* July 9. http://www.thecanadianencyclopedia.ca/en/article/inuktitut-words-for-snow-and-ice/ (accessed Sep 12, 2015).

---

[12]Valdes-Dapena (2013).

[13]Pearson (2013).

The Economist. (2016). The collaboration curse: The fashion for making employees collaborate has gone too far. *The Economist.* Jan 23. http://www.economist.com/news/business/21688872-fashion-making-employees-collaborate-has-gone-too-far-collaboration-curse (accessed Jan 25, 2016).

Valdes-Dapena, V. (2013). 870 k Toyotas recalled for spider-related problem. *CNN Money.* Oct 17, 2013. http://money.cnn.com/2013/10/17/autos/toyota-spiders-airbag-recall/index.html (accessed Oct 29, 2013).

Varjo, J., Hongisto, V., Haapakangas, A., Maula, H., Koskela, H., & Hyönä, J. (2015). Simultaneous effects of irrelevant speech, temperature and ventilation rate on performance and satisfaction in open-plan offices. *Journal of Environmental Psychology, 44,* 16–33.

Social Psychology

Abstract From the standpoint of Herbert Simon's model of rational design, good design is a matter of expertise. That is, the more expert designers are, the more closely they approximate the ideal of rational design. There are many forms of expertise. One kind of expertise that is relevant to social aspects of good design may be found in social psychology. Roughly speaking, social psychology concerns how people's thoughts, feelings, and actions are affected by other people around them. Many popular online services, such as Facebook, are designed expressly to exploit people's openness to influences from others. Two concepts relating to social psychology and design are introduced and discussed here, namely technotonicity and trust. Technotonicity concerns how designs make people feel, especially in how they shape people's interactions with each other. Trust is a social attitude that describes not only how people feel they can interact with each other but also how they feel they can interact with their goods. Technotonicity and trust suggest the importance of social expertise to the issue of good design.

Social Expertise

On Herbert Simon's view, expertise is key to good design. Thus, the more knowledge people have, the better they are able to assess designs. From a technology-society perspective, this means increasing social expertise, that is, knowledge of how people conduct their social lives.

One area where relevant expertise may be found is social psychology. Roughly speaking, social psychology is the study of how people's thoughts, feelings, and behavior are influenced by the presence of other people, whether actual or imagined.

A simple example would be peer pressure. Everyone can probably remember a time when they have done something they would not otherwise do because of the urging of peers. Conversely, who has not been restrained from doing something they wanted to do by the disapproval of others?

Knowledge about peer pressure can be instructive in understanding design. Jaron Lanier notes how the Beacon feature of Facebook was designed to use peer pressure

© Springer International Publishing AG 2017

C. Shelley, *Design and Society: Social Issues in Technological Design*,
Studies in Applied Philosophy, Epistemology and Rational Ethics 36,
DOI 10.1007/978-3-319-52515-0_3

to sell things.[1] With Beacon, any purchase made through a Facebook partner was broadcast to the purchaser's Facebook friends. These messages, in turn, served to make those products look more popular to those friends who might, then, be persuaded to purchase them as well. However, the feature raised privacy concerns and was subsequently withdrawn.

Research has also shown that peer pressure can be harnessed to help people to achieve their life goals. People can be motivated to persist in programs to achieve life goals such as fitness, cleanliness, or professional achievement when encouraged by approval from friends on social media platforms. Designers can sometimes get people to work out for likes on Facebook![2]

In this chapter, concepts from social psychology are examined that provide information relevant to good design. In particular, we will examine the concepts of technotonicity and trust. Many more concepts could be included but these two provide a useful introduction to the importance of social psychology in assessment of good design.

Case Study: Revolving Doors

The inventor Theophilus Van Kannel (1841–1919) reportedly disliked conventional doors.[3] In particular, he disliked the social problem of knowing when to hold a door open for other people to use. ("After you. No, after you".) Being a gentleman, for example, demands men hold doors open for women. However, it can be inconvenient for the person holding the door, and unnecessary for the person for whom the door is being held open.

Q: When do you hold a door open? In what way? Why, or why not?

Being an inventor, Van Kannel set out to solve the problem with a new door design. The result was the first revolving door. In a revolving door, three or four doors rotate around a central, vertical shaft within a cylindrical enclosure. When users push on one door, the whole assembly spins, allowing them to enter or exit (Fig. 1).

Revolving doors do solve Van Kannel's social problem: They cannot be held open, thus obviating the need to hold them open.

In addition, revolving doors have a number of other advantages. The doorway can be kept sealed, acting as an airlock. This arrangement helps to slow heat flow in or out of the building and prevents exposure to wind gusts near the doorway. Similarly, revolving doors keep street noises and smells out of buildings. Because

---

[1] Lanier (2010), p. 54.
[2] Hamari and Koivisto (2015).
[3] 99% Invisible (2013).

of the seal, revolving doors are often easier to open than swinging doors, which can become immobilized by pressure differences.

Architects sometimes employ big revolving doors to create dramatic entrances to big buildings.

Van Kannel started the successful Van Kannel Revolving Door Co. He was also awarded the John Scott Medal by the Franklin Institute of Philadelphia in 1889 in recognition of the significance of this invention.

However, revolving doors can lead to uncomfortable situations.

> Q: Do you use revolving doors? What difficulties do people encounter using them?

One difficulty is that revolving doors can become jammed, stranding people in a compartment. In the movie *The Godfather*, Carmine Cuneo is shot to death after being trapped in a revolving door compartment by a gunman from another crime family.[4]

For these reasons, it seems, revolving doors are typically used only about 25% of the time when a swinging door is available at an entrance.[5]

Technotonicity: Technotonic

One of the basic facts about any design that people have to deal with is stress. Van Kannel found regular doors stressful, that is, they gave rise to social situations that made him anxious. His solution was to design a door that eliminated the causes of that stress. However, the new revolving door design introduces new sources of potential stress.

To understand this issue more clearly, we may start with the concept of *technotonicity*. Technology scholar Ron Westrum discusses reasons why people's responses to designs may be emotionally positive or negative.[6] A design that provides an emotionally positive response is *technotonic* whereas a design that evokes a negative response is *technostressing*.

Let us begin with the concept of technotonic design. Westrum defines this as follows: A design is technotonic to the extent that it is pleasurable or reinforcing to the user in the sense that

- It gives users a feeling of control or mastery over the environment.
- Its use reflects a high degree of skill.

---

[4]Coppola (1972).
[5]Cullum et al. (2006).
[6]Westrum (1991), pp. 221–223.

Fig. 1 International revolving door in Turkey. The door adds emphasis to the building entrance. Photo by Moberg/Wikimedia commons. Detail of URL: https://commons.wikimedia.org/wiki/File: Revolving_Door.jpg#/media/File:Revolving_Door.jpg

- Its appearance evokes aesthetic pleasure.
- It evokes pleasant associations.

Case Study: The Bugaboo Frog

A good example might be the modern baby stroller. Busch notes that items like the Bugaboo Frog baby stroller, from Dutch designer Max Berenberg, are very plea-surable for parents (Fig. 2).[7] First, the stroller is quite flexible (p. 28):

> Lightweight and with a seat that adjusts to three positions, it has two small swiveling wheels for city maneuvering, and two large terrain wheels for off road. It can convert to a two-wheel position to be pulled on the beach or snow, has a reversible handlebar and a seat/bassinet, enabling the child to face either direction.

This feature affords the user great control over the function of the stroller and allows them to take it into many different environments at their convenience.

---

[7]Busch (2004), pp. 27–34.

Second, all this flexibility provides users with opportunities to achieve a mastery of the stroller's different configurations. Indeed, Berenberg complains that people often do not make use of all the functionality of the stroller.[8]

However, Busch quotes a review from the New York Times in 2003 in which the reviewer notes how he enjoys the superior handling abilities of the stroller[9]:

> Maybe it was the 12-inch all-terrain tires or the squishy grip bar or the fact that the Bugaboo steered more like a Porsche than a pram, but there I was, wheeling an empty stroller through a grocery store for the adrenaline rush.

This handling quality allows the user of the stroller to feel skillful in comparison to users of clunkier designs.

Third, the appearance of the Bugaboo appeals to modern sensibilities (p. 28):

> And its styling reflects its contemporary functionality—with black rubber and fabric seat that comes in red, gray, or aubergine, it is solid, streamlined, contemporary. There is nothing frivolous about this baby accessory.

These qualities are ones that modern consumers tend to enjoy.

Fourth, the Frog was designed to appeal to men. Another aspect of modern life is a greater role for fathers in child care. As a result, the Bugaboo employs a design language that men might find familiar and agreeable (p. 30):

> With dads taking a greater role in childcare, small wonder that oversized SUVs have become the new model for strollers—though it goes without saying that like most SUVs, many of these strollers are likely to be used more in the suburban avenues of malls and food courts than in any more demanding rural terrain.

In short, the Bugaboo evokes associations with large, manly vehicles that its male clientele may find pleasant and reassuringly masculine.

Q: What designs do you find technotonic? Why?

Bugaboo strollers are so well-known in the Netherlands, where the company is headquartered, that they were celebrated on a Dutch stamp in 2007.[10]

Technotonicity: Technostressing

Of course, if some designs evoke pleasant feelings, then other designs may evoke unpleasant ones. For such designs, Westrum applies the term *technostressing*. A design is technostressing to the extent that it causes stress in the sense that

---

[8]Carter (2013).

[9]Cf. Hochman (2003).

[10]Fairs (2007).

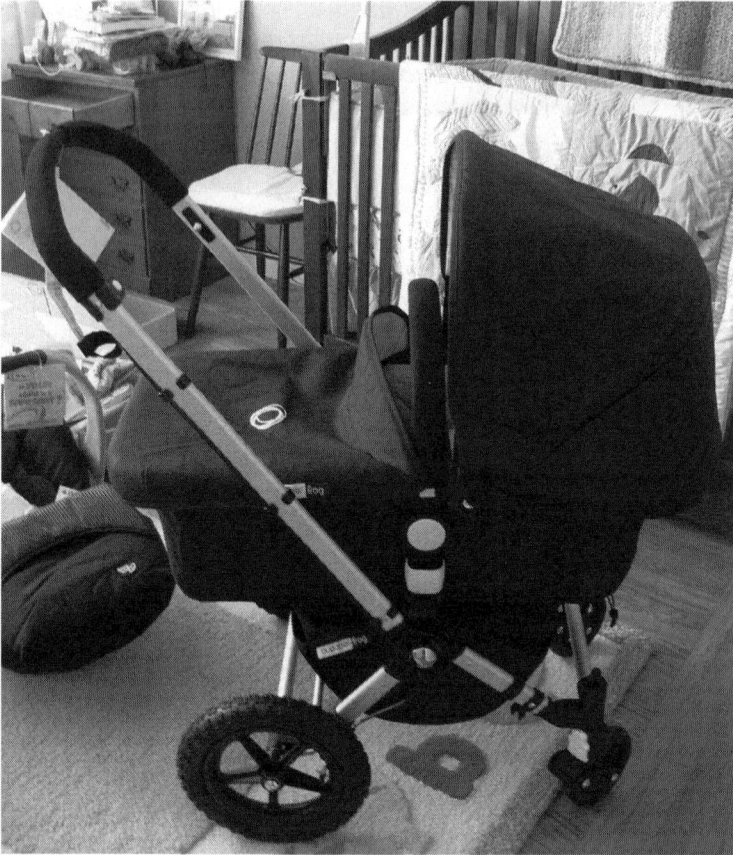

Fig. 2 A Bugaboo Frog stroller. Photo by Jessica Merz. URL: https://flic.kr/p/nPYJJ

- It removes feelings of control or mastery from users.
- It demonstrates users' lack of skill and knowledge of the device.
- It is ugly or evokes bad sensations.
- It achieves disrepute through association.

Case Study: Digital Watches

A classic example of a technostressing design would be the common digital wristwatch or, more precisely, the controls of digital wristwatches (Fig. 3). Controls on a typical digital wristwatch amount to three or four small buttons protruding from its sides. To use a digital watch, users must program it by pushing the little buttons in different combinations. There is little rhyme or reason in the correspondence between combinations and their results. Watches usually come with a manual that is normally lost right away, leaving users with nothing but the buttons when it comes time to reprogram them.

Here is an illustrative story by designer Bill Moggridge, founder of the design firm IDEO, about a digital wristwatch that he bought for his young son[11]:

Everything went well for six weeks. He didn't seem to miss the radio (poor reception), used the alarm every morning, and enjoyed the cool design. Then two things happened; daylight saving time ended, and he gave up his paper route. He left the watch on the chest of drawers in our bedroom, as he said, 'Could you cancel the alarm and change the time, please, Dad?'

By that time of course the instructions were lost, so I tried to make the adjustments by pushing buttons in a vague and unstructured way, hoping that some automatic memory would make me get it right.

My wife is a lighter sleeper than I, so it was she who got out of bed to cancel the alarm when it went off at four o'clock in the morning, an hour earlier after the time change. I tried to reset it again the next night, but with the same result. She was starting to get irritated, so the following night I took the battery out, and assured her that an undisturbed night would follow. At four o'clock the next morning, there it was again, 'Beep-be-be-be-beep,' in an ascending volume and persistent shrill tone. That was too much! She woke me up, marched out of the bedroom and returned in a moment with a hammer. That was the end of the watch! It turned out that the battery that I had removed was for the radio; there was another one buried deep inside that powered the watch.

It takes little thought to see how this digital watch is technostressing. First, the watch did not provide appropriate feedback to Moggridge's efforts at reprogramming it. Pushing the buttons evidently led him to have only a vague hope that it would perform as desired—a false hope as it turned out. Second, the watch controls were cryptic and arbitrary, so that no expertise already possessed by Moggeridge could be applied to it. Put another way, the watch controls were unintuitive and, due to lack of good feedback, difficult to learn. It is the sort of design that makes users feel stupid. Third, although the watch had "cool design", it produced a shrill alarm tone that could not be deactivated, which ultimately led to its destruction.

It is not clear that digital watches have any bad associations. In the past, specific wristwatch designs such as calculator wristwatches have had nerdy associations, which was widely seen as negative. In future, if smartwatches become more common, then plain digital wristwatches may be seen as outdated or low status.

Q: What designs do you find technostressing? Why?

It may seem that technotonic design is always good design and technostressing always bad. However, it may be appropriate for designs to be technostressing. For example, the watch alarm described by Moggeridge was clearly technostressing.

---

[11]Moggeridge (2007), pp. 4–5.

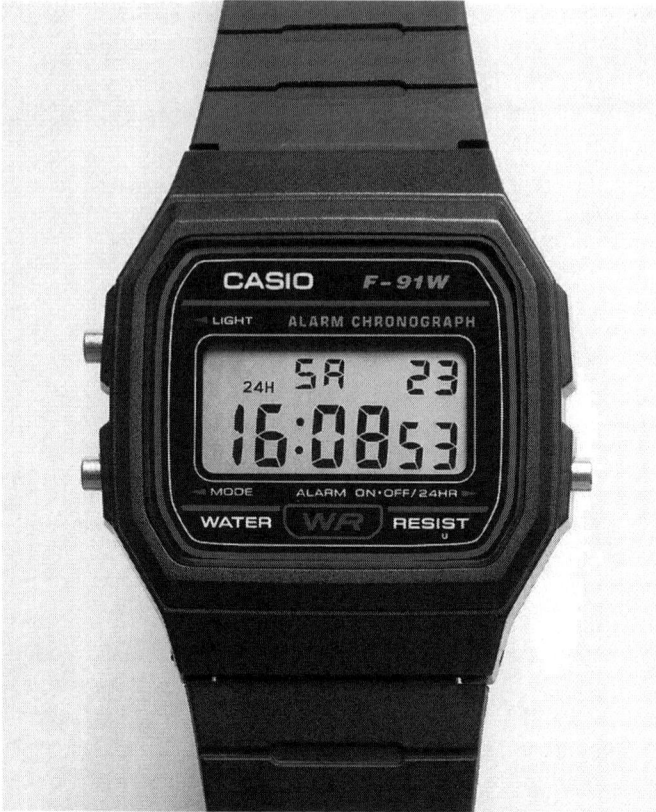

Fig. 3 Casio F-91 W digital watch. Photo by Petar Milošević/Wikimedia commons. URL: https://
commons.wikimedia.org/wiki/File:Casio_F-91W_digital_watch.jpg#/media/File:Casio_F-91W_
digital_watch.jpg

Yet, such an alarm might have to be irritating if it is to fulfill its function of waking up sleepy users.

For another example, consider a project undertaken in the German city of Hamburg to fend off men who pee on walls in the St. Pauli district, a hazard due to the many bars and clubs in the area.[12] Walls in the district were coated with a superhydrophobic paint, that is, paint that violently repels water. People urinating on such a wall would be copiously splashed with their own urine. That experience is certainly technostressing! The aim of the design is expressly to make public urination unpleasant and, therefore, to encourage use of better facilities.

The point is that either technotonic or technostressing design may be employed depending on which approach is optimal for obtaining the desired result.

---

[12]O'Sullivan (2015).

Social Influence

Westrum's concepts are instructive but lack an explicitly social element. That is, the response of users of a design is envisioned in isolation from the social environment. This situation may be envisioned as in Fig. 4.

However, social psychology suggests that a person's response to a design may be affected in a social situation, that is, when other people are present (or imagined). This situation may be envisioned as in Fig. 5.

The presence of other people can profoundly influence how users experience a design. As a student told me once, it is irritating when a balky bus door will not open to let you off a bus. However, it is embarrassing when a bus full of people are observing you struggle with it.

Fig. 4 A user responds to a design in isolation

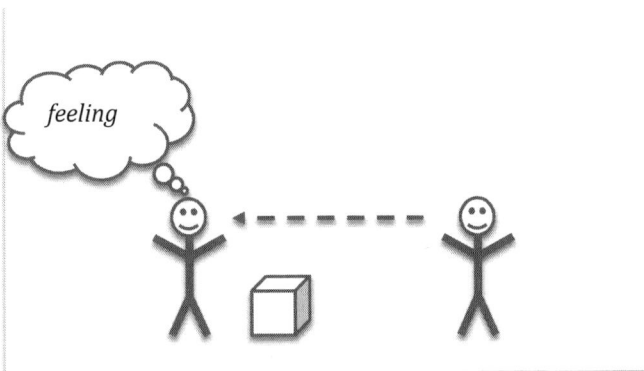

Fig. 5 A user responds to a design when other people are present

Fig. 6 Google executive Amanda Rosenberg modeling the Google Glass face mounted wearable computer. Photo by Max Braun/Wikimedia commons. URL: https://commons.wikimedia.org/wiki/File:Google_executive_Amanda_Rosenberg_modeling_the_Google_Glass_face_mounted_wearable_computer.jpg

Or, consider the failure of Google Glass in the marketplace (Fig. 6). Although the design was technically successful and won some people's approval, too many people found their experience with it to be technostressing[13]:

> Even some of the early adopters are getting weary of the device. "I found that it was not very useful for very much, and it tended to disturb people around me that I have this thing," says James Katz, the director of emerging media studies at Boston University's College of Communication.

In other words, other people's responses to Google Glass affected how this user felt about it. Clearly, this sort of consideration is important in the evaluation of good design.

Social Technotonicity: Technotonic

To account for the importance of social psychology in people's responses to designs, we may extend Westrum's concepts explicitly into social situations.

A design is *socially technotonic* to the extent that its use

- Promotes harmonious social interactions, or
- Enhances users' social standing, or
- Increases users' attachment to their social group.

Consider the Bugaboo Frog stroller again. Features of this design tend to make it not only technotonic but also socially technotonic. First, strollers make it

---

[13]Metz (2014).

straightforward for parents to take their children and supplies with them into public, where they can meet and make friends with the children of other parents.

Second, Busch notes that the stroller can be a "status object".[14] He speaks at length about how the design of the stroller conveys a sense of safety to parents who use it. The sun and rain shields it provides, the shock absorbers, big tires, and titanium frame, like the sturdy construction of SUVs, reinforce the idea that small children are constantly in danger from the outside world (p. 31):

> The real service of strollers, [parents] say, is that while offering parental convenience (many must also double as shopping baskets and storage bins), they provide a small, safe place for children to sit in a chaotic and unpredictable world.

By purchasing and using a stroller so obviously designed to satisfy this (perceived) need for security, parents see themselves as doing their social duty, and are seen by others in the same light. These users then feel good about themselves.

Third, use of a particular design of stroller can be seen as a sign of adherence to a particular social group. Carter describes the user of a Bugaboo Frog as "the archetypical yummy mummy, clad in Brora and pushing her Bonton-clad tot to the yoga class."[15] In short, people perceive users of Bugaboo strollers as belonging to a certain social group.

Carter also notes that sales took off after movie star Gwyneth Paltrow was seen using one in 2004 in New York City. So, anyone wanting to join the ranks of Gwyneth Paltrow, Gwen Stefani, Hugh Jackman, or Kate Middleton can do so by purchasing a Bugaboo, if they can afford it.

Social Technotonicity: Technostressing

By contrast, a design is *socially technostressing* to the extent that its use

- Promotes hostile or embarrassing social interactions, or
- Discredits users, or
- Detaches users from their social group.

A typical parking lot is a good example of a design that is socially technostressing (Fig. 7). First, parking lots tend to place drivers and pedestrians in the same space, sometimes leading to unpleasant conflicts. When cars are crowded together, there is also the possibility of disputes over who has the right to park in a given spot, giving rise to angry confrontations known as *parking rage*.[16] Also, parking lots provide hiding places for potential assailants that may make vulnerable people reluctant to use them, particularly at night.

---

[14]Busch (2004).

[15]Carter (2013).

[16]Cf. (Grove et al. 2004).

Fig. 7 Mall parking lot. Photo by Daniel Oines. URL: https://flic.kr/p/bap9e2

Second, one location in a parking lot is often hard to distinguish from another. Thus, it is easy for people to forget where they parked their cars. Having to wander through a parking lot in search of a car can be embarrassing to the searcher.

Third, parking lots single out certain people for special attention. For example, parking places for disabled people are often provided. While the convenience may be appreciated, the conspicuous signage may make users feel that unwanted attention is being drawn to their personal situation.

Q: What things to you find socially technotonic? Socially technostressing?

Trust

We have discussed how people's responses to designs may be conditioned by the presence of others. However, such responses go beyond being merely positive or negative. Experiences of designs can evoke complex social attitudes. An important example is *trust*. It is important in social life to know *whom* to trust. In a world of complex technology, it is also important to know *what* to trust. Thus, knowledge of trust is also important for designers from a technology-society perspective.

Although trust is a complex topic, the essentials are captured in this characterization by Lee and See[17] (p. 54):

Trust "is the attitude that an agent will help achieve an individual's goals in a situation characterized by uncertainty and vulnerability."

This definition identifies three factors that apply to situations where trust is involved. First, trust involves needing (or wanting) the assistance of other people in order to attain goals. Second, there is uncertainty about whether or not the other people in question are willing to give assistance. Third, if the other people do not assist, then there will be some cost to the trusting person as a result.

Trust is important for people because it helps them to achieve things together. For example, when students work in groups on an assignment or project, they need to find some way of arranging things so that they can trust other members of their group.

In some groups, there is a high level of trust, so that members simply rely on the others to do their parts. In some groups, there is a low level of trust, so that members have to spend a fair amount of time managing the work of the other members.

Trusting Things

As noted above, trust applies not only to people but to things. Consider this example of a car-owner, Dave, who wrote to a newspaper car columnist to ask whether or not he should trust the oil-change light in his Honda Civic[18]:

The Honda has an oil life feature that tells me what percentage of oil is left. Once it gets below 20% a little wrench lights up telling me to change it. Initially I thought this system was actually monitoring the oil and advising me of its condition. I've since learned that the system has to be reset after each oil change, making me wonder if it is just hooked into the odometer. Should change oil according to the driving conditions?

The issue of trust in play here is similar to that for people. It concerns what attitude Dave should take towards his car, specifically the design of the oil monitoring system, represented by the little wrench light on his dashboard. Dave's goal is not made explicit but we may assume that he wants to keep his car in good working order. The issue in his mind is whether or not the designers of the car's oil monitoring system share that goal. Having been told that the oil monitor lights up very time his car is driven more than a certain distance, he is uncertain about the goals of the designers. Perhaps they want him to change the oil more often than is strictly necessary. Dave is vulnerable in the sense that too many oil changes will cost him money without benefit, while too few changes will wear out his car's engine prematurely.

---

[17]Lee and See (2004).
[18]MacGregor (2012).

From this example, it appears that the same definition of trust that applies to people also applies to designs that they use.

The automotive columnist who answered the letter recommended that Dave should trust his car's indicator light. He explained that oil indicator depends not only upon mileage but also engine revolutions, the number of times the car is started, how long the engine runs between stops, ambient temperatures, coolant temperatures, engine loading, and other parameters. On this basis, he urged Dave to "follow the advice given to you by your car."

Note how the explanation addresses the issue of trust. It attempts to remove crucial uncertainties from the situation by explaining how the oil monitoring system works. In so doing, it also makes a case that the interests of the car designers are consistent, and not at odds with, the interests of the owner. By correcting the owner's misbeliefs about the cars designers, the columnist aims to re-establish Dave's trust in their work.

Level of Trust

As noted above, trust comes in different levels and it is important to know how much to trust others. The same consideration applies to technology. So, we face the problem of how to calibrate trust appropriately to the situation. Lee and See note that, crudely speaking, there are three possibilities:

1. Appropriate: trusting something to the right extent;
2. Over-trust: trusting something too much; and
3. Under-trust (distrust): trusting something too little.

The case of the Honda Civic owner above illustrates under-trust. Dave placed less trust in the oil monitoring system than was appropriate. Under-trust is potentially bad because users of the technology may not derive full benefit from it. Distrusting the wrench light, Dave may change the oil too infrequently and thus subject his car to damaging wear and tear.

Over-trust is potentially bad because the user of the technology could be placed in a vulnerable position by it. For example, too much trust in the unsinkability of the Titanic caused the operators to provide too few lifeboats for a safe evacuation in the event of disaster.

Q: What designs have been trusted inappropriately? Why?

Perhaps the most famous example of over-trust was the Titanic. The ship was equipped with enough lifeboats to evacuate only about half of the passengers and crew that it could carry. Regulators at the time approved of the arrangement. It was assumed that the Titanic was practically unsinkable. Thus, in case of distress, only a small number of lifeboats would be needed to ferry people from her to rescue ships over several trips.[19]

Assessing Trustworthiness

As trust is a complex, social attitude, there is no simple way to assess whether or not a design communicates an appropriate level of trust to people. A revolving door may be very reliable, for example, but people will shun it if they remain afraid of getting trapped in its compartments.

Trust has become an important factor in the design of web- or app-based services.[20] On Ebay, for example, people are often put in a position of purchasing items from others whom they do not know at all. This situation creates a problem of trust because of this uncertainty and the fact that the seller might lie about what they are selling or simply keep the purchase money and not send the item that was bought. To deal with this issue, it is crucial for Ebay to provide some substitutes for a trusting relationship. In this case, the substitutes include a rating system whereby buyers and sellers rate each other. In this way, potential buyers can rely on the experience of others to gain a sense of how trustworthy sellers are. Another substitute is a dispute-settlement mechanism whereby Ebay will help to resolve any disagreements between buyers and sellers. This provides each party with some insurance in the event of misbehavior by the other.

Case Study: Dog Sitting

For another example, consider an American web-based service named Rover.com. This service allows people who do not want to use kennels to hire people to sit their dogs while they are away. As with any contracting service, trust is an important issue for Rover.

Q: In what ways does trust figure in hiring a dog sitter?

Rover has implemented a number of measures to help people to trust the service.

1. The service carries premium pet insurance and has 24/7 veterinarian consultations.

---

[19]Berg (2012).
[20]Andruss (2015).

2. The service allows sitters to provide photos and videos of them interacting with their pet "guest".
3. The service includes client ratings of dog sitters.
4. Sitters can also provide "trust criteria" such as third-party references, background checks, or work history.
5. The company has a RoverCam that shows "all the action at the dog-friendly company headquarters."

In spite of these features, not everyone considers the service trustworthy. Jodi Hamilton, of Fort Wayne Animal Care and Control, notes that Rover.com does not perform any background checks of its own. She also worries about how well sitters would monitor interactions with their own pets, or take precautions to prevent the spread of diseases. In addition, pet-sitting services in private homes may be considered illegal.[21]

> Q: Would you consider Rover.com trustworthy?

How people use and respond to designs depends in part on the social situation they are in. As such, knowledge of how social situations affect people's thinking and behavior allow us to make better assessments of designs. Concepts such as technotonicity and trust illustrate how knowledge of social psychology can help with this task.

It might seem that good design is always a matter of relieving stress and increasing trust. Sometimes, that is the case. However, it is sometimes appropriate for designs to cause stress or provoke distrust where these results are best in order for designs to achieve their ends.

References

99% Invisible. (2013, Nov 7). *Why don't people use revolving doors?* Retrieved Nov 9, 2013, from Slate: http://www.slate.com/blogs/the_eye/2013/11/07/revolving_doors_why_don_t_we_use_them_more.html.

Andruss, P. (2015, Jan 10). *What to consider before launching a business in the sharing economy.* Retrieved Feb 20, 2015, from The Globe and Mail: http://www.theglobeandmail.com/report-on-business/small-business/sb-growth/day-to-day/how-to-launch-a-business-in-the-sharing-economy/article22226642/.

Berg, C. (2012, Apr 12). *The real reason for the tragedy of the Titanic.* Retrieved Aug 17, 2016, from The Wall Street Journal: http://www.wsj.com/articles/SB10001424052702304444604577337923643095442.

Busch, A. (2004). *The uncommon life of common objects: Essays on design and the everyday.* New York: Bellerophon.

---

[21]Kilbane (2015).

Carter, K. (2013, May 23). *Bugaboo designer Max Barenbrug: the master of reinvention.* Retrieved Sept 23, 2015, from The Guardian: http://www.theguardian.com/lifeandstyle/2013/may/23/bugaboo-designer-max-barenbrug-buggy.

Coppola, F. F. (Director). (1972). *The Godfather* [Motion Picture].

Cullum, B. A., Lee, O., Sukkasi, S., & Wesolowsky, D. (2006). *Modifying habits towards sustainability: a study of revolving door usage on the MIT campus.* Cambridge: Massachusetts Institute of Technology.

Fairs, M. (2007, Jan 6). *Dutch design stamps issued.* Retrieved Sep 6, 2016, from Dezeen: http://www.dezeen.com/2007/01/06/356/.

Grove, S. J., Fisk, R. P., & John, J. (2004). Surviving in the age of rage. *Marketing Management, 13*(2), 41–47.

Hamari, J., & Koivisto, J. (2015). Working out for likes: An empirical study on social influence on exercise gamification. *Computers in human behavior, 50*, 333–347.

Hochman, D. (2003, Sep 7). *Baby's latest power trip: Sport utility strolling.* Retrieved from The New York Times: http://www.nytimes.com/2003/09/07/style/noticed-baby-s-latest-power-trip-sport-utility-strolling.html.

Kilbane, K. (2015, Sep 22). *Rover.com, an Uber-like version of pet sitting, raises some concerns in Fort Wayne.* Retrieved Jan 12, 2016, from News-Sentinel.com: http://news-sentinel.com/news/local/Rover-com–an-Uber-like-version-of-pet-sitting–raises-some-concerns-in-Fort-Wayne.

Lanier, J. (2010). *You are not a gadget.* New York: Alfred A. Knopf.

Lee, J. D., & See, K. A. (2004). Trust in automation: Designing for appropriate reliance. *Human factors, 46*(1), 50–80.

MacGregor, R. (2012, Sep 12). *Trust your car when it asks for an oil change.* Retrieved Sep 20, 2012, from The Globe and Mail: http://www.theglobeandmail.com/globe-drive/culture/commuting/trust-your-car-when-it-asks-for-an-oil-change/article1359853/.

Metz, R. (2014, Nov 26). *Google Glass Is dead; long live smart glasses.* Retrieved Nov 27, 2014, from MIT Technology Review: http://www.technologyreview.com/featuredstory/532691/google-glass-is-dead-long-live-smart-glasses/.

Moggeridge, W. (2007). *Designing interactions.* Cambridge, MA: MIT Press.

O'Sullivan, F. (2015, Mar 4). *A novel solution to public urination: Walls that splash pee right back at you.* Retrieved Mar 5, 2015, from The Atlantic: Citylab: http://www.citylab.com/design/2015/03/a-novel-solution-to-public-urination-walls-that-splash-pee-right-back-at-you/386791/.

Westrum, R. (1991). *Technology: The shaping of people, things, and society.* Belmont, CA: Wadsworth.

Style

Abstract Almost any discussion of good design includes considerations of style. Style often refers to the relationship between the appearance or form of a design and its function. There is much controversy over what constitutes good style. A reductionist view is that good style is just the same as good design for function. That is, anything that works well will thereby exhibit good style. Such designs are often described as elegant or honest. Dieter Rams supports this view. A structuralist view of good style is that the appearance of a design may be used to accomplish social as well as functional goals. A pair of glasses, for example, may be used both to correct vision problems and to convey membership in a social group such as hipsters. Such designs may be said to possess decorum. A commercialist view is that good style is a way of marketing designs. The appearance of a design should evoke associations with its originators in the minds of consumers, thus helping its originators to market it effectively. Consumers immediately associate the form of a Coca Cola bottle with the Coca Cola Company, for example. This sort of association is often known as branding.

Form and Function

Many would agree that style is an important consideration in good design. However, the term often means different things to different people. In one sense, style refers to concepts such as *individual style* or *design style*, which refers to how designs reflect the affiliations of their originators.[1] In architecture, for example, a building might be described as *modernist* because its appearance includes elements that are associated with the modernist movement in architecture.

In another sense, *style* refers to the relationship between the appearance and function of designs. Sometimes, the term is used in this way to praise a design, as when people describe a design as elegant to mean that it both looks good and works well, typically on the same grounds. However, the term is also used as a form of

---

[1]Cf. Chan (2015).

© Springer International Publishing AG 2017

C. Shelley, *Design and Society: Social Issues in Technological Design*,
Studies in Applied Philosophy, Epistemology and Rational Ethics 36,
DOI 10.1007/978-3-319-52515-0_4

abuse, as when people describe a design as all style and no substance, implying that its good looks disguise its lack of functionality.

Because of its relationship to assessment of designs, the second sense is the one explored here.

There are many views about how style or attention to appearance relates to matters of good design. As we have seen, Dieter Rams thought that appearances matter insofar as they help a design to fulfill its purpose. By the same token, any design in which appearances mislead users or interfere with functionality is a bad one.

Yet, such a view may be overly narrow. Functionality is sometimes defined in a way that excludes any social significance of appearances. Recall that Dieter Rams urged designers to regard their works as tools, like hammers. Yet, tools are typically locked away out of sight when not in use. However, people sometimes want to put their things on display for others to see, perhaps to show off. It is not self-evident that facilitating such displays represents bad design.

The social significance of appearances also provides opportunities for product marketing through style. Marketers use style to establish their products as brands in the marketplace. They ensure that any designs they sell reflect their corporate identity, e.g., the characteristic shape of a Coke bottle. They use these character-istics to build up a positive impression among consumers in order to increase consumer loyalty. In marketing, good style means stimulating sales.

In this chapter, we will examine three positions about how style relates to good design. The first position is *reductionist*, on which good style is simply a matter of good function and nothing further. The second position is *structuralist*, on which good style is a matter of placing designs appropriately in the prevailing social framework. The third position is *commercialist*, on which good style is a matter of good marketing.

Reductionism: Style is Function

Dieter Rams holds that good design is beautiful but he takes a *reductionist* view of what that means. Consider number 3 of the Ten Commandments of Good Design:

> Good design is aesthetic[2]: The aesthetic quality of a product is integral to its usefulness because products we use every day affect our person and our well-being. But only well-executed objects can be beautiful.

Here, Rams allows that aesthetics is important to good design. He denies, however, that achieving good-looking design involves anything more than achieving a design that works well and obeys the other commandments of good

---

[2]Lovell (2011), p. 354.

design. In other words, he reduces the problem of making designs that are stylish to the problem of making designs that function well.

This view explains why Rams complains about designers who try to add "fireworks of signals" in their works to wow clients with their cleverness or good taste.[3] Style is not something that is added to a design. Instead, it is integral to the achievement of functional design.

Rams's view is not unprecedented and could be compared with a longstanding view that the best things look best when unadorned. The classical Greeks, for example, held that the bodies of top athletes were the best human bodies.[4] To dress them up would merely diminish their appearance. So, Greek athletes competed in the nude.

Q: Is Rams correct?

Honesty

The idea that there is something confusing or even deceitful about setting out to make designs that are stylish as such is found in another form in the fifth of Rams's Ten Commandments of Good Design:

> Good design is honest[5]: It does not make a product more innovative, powerful or valuable than it really is. It does not attempt to manipulate the consumer with promises that cannot be kept.

In other words, a design should not be made to appear to be something that it is not. Otherwise, users could be deceived about what a design can deliver, in which case their decision to purchase and employ it is made under false pretenses.

In fact, honesty in design is a principle that goes back earlier than the modernists to the Victorian architect A.W.N. Pugin.[6] It may be defined briefly as follows. An honest design:

1. Does not disguise what it is, and
2. Exhibits what it is.

Consider the design of calculator apps. Apple provides a calculator app (or "desktop accessory" in earlier versions) with its operating system software. Since the beginning, the graphical interface to this app has used shadows and shading to make it appear that the calculator and its buttons are 3-dimensional objects (Fig. 1).

---

[3]Rams (1984), p. 112.
[4]Gardiner (1930), p. 58.
[5]Lovell (2011), p. 354.
[6]Conway and Roenisch (2006).

Fig. 1 The iOS 3.1
calculator app. Photo by
Dominic Alves. **URL:** https://
flic.kr/p/6Y7apN

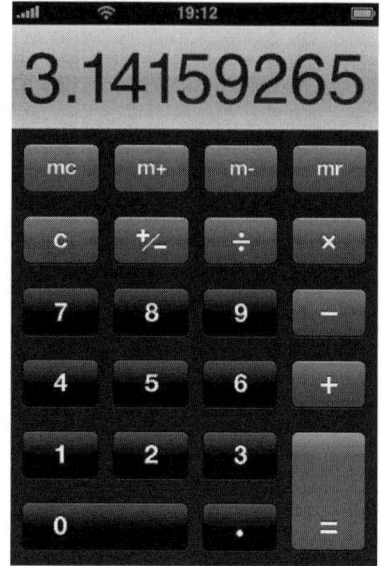

The app made it appear as though there were calculator buttons protruding out of the screen, which was not actually the case. This was done because it made the app look more like a real calculator, with which users were already familiar. However, this design is dishonest: The interface is really 2-dimensional.

Recently, there has been a trend towards "flattening" of interface designs. That is, pseudo-3-dimensional interfaces have been replaced by obviously 2-dimensional ones. The use of 2-dimensional interfaces is honest in the sense that it recognizes that the computer (or phone or tablet) screen provides a 2-dimensional surface to users (Fig. 2).[7]

Dishonesty

On occasion, designers decide to employ dishonest designs. For example, some cell phone towers in warm climates are disguised to look like palm trees (Fig. 3). The idea is clearly to take a design that is often considered ugly—the standard cell phone tower —and make it resemble something that people that people find visually acceptable.[8]

The ruse may be effective in the sense that people are less apt to complain about the appearance of disguised cell phone towers than about undisguised ones. Nevertheless, designers like Dieter Rams might argue that such a move is just lazy. That is, instead of working hard to create a design that is properly functional and

---

[7]Bilton (2013).
[8]Byrnes (2013).

Fig. 2 The iOS 9.3
calculator app. Photo by
Cameron Shelley

therefore good looking, designers simply create something that is cheap and works
well enough and then apply a fake exterior in order to cover up their sloppiness.

Q: Can you think of other honest or dishonest designs? Are they good
designs?

For modernist designers, like Rams, honesty is a way of keeping aesthetics in its
proper place, that is, as just an aspect of functionality. To elevate it beyond this role
would be counterproductive, at best, and uncivilized, at worst.

Structuralism: Style is Social Signaling

Rams makes a forceful argument: Designers should not mix design with art. Get the
design right, and the beauty takes care of itself.

However, it may also be the case that there are relevant points that Rams has not
considered. He assumes that aesthetic appeal, as such, has no utility for people. Of
course, they like it but it does nothing to improve their lives. Perhaps this premise is
wrong. An alternative view would be that aesthetics has a special utility for people,
a social utility.

Fig. 3 Cell phone tower
disguised as a palm tree.
Photo by
Graphicsclz/Wikimedia
commons. URL: https://
commons.wikimedia.org/
wiki/File:Palma_DSC02769a.
jpg

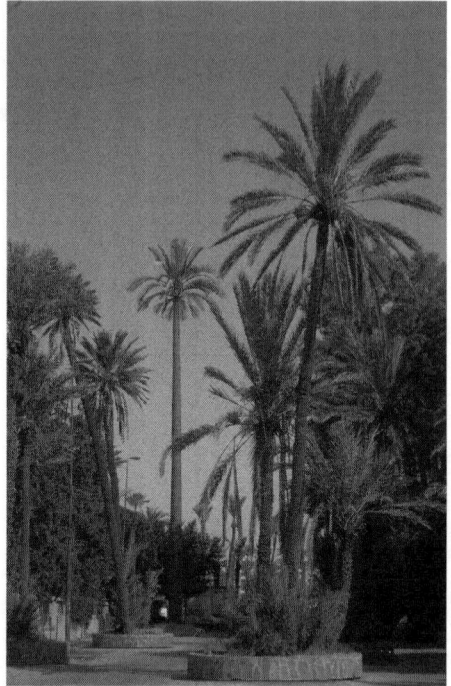

This idea is central to the *structuralist* view of style. From the structuralist view, people use design to accomplish social goals, such as showing where they fit within the prevailing social structure. For example, eyeglasses all have the same basic function of correcting for defects in vision. Insofar as function is concerned, then, eyeglasses should all look more or less alike. However, they often look quite different. Opticians may carry hundreds of different frames in their inventories.

The reason for the diversity of eyeglass styles is that people use these differences to tell other people about themselves. Someone who wears mirrorshades, for instance, may do so to signal to others they are cool and detached from the world. A hipster may use blocky retro-glasses to suggest an ironic outlook on life. So, glasses can communicate more than "the wearer has vision problems" or "it is sunny out."

Style allows designers to provide for this social need (Fig. 4).

Decorum

A traditional term for the use of style for social signaling is *decorum*. Decorum refers to a concept of good design in architecture first stated by the Roman architect Vitruvius and which persisted in various forms into the nineteenth century. In rough

Fig. 4 What signal do these sunglasses convey? Photo by songjayjay. URL: https://pixabay.com/en/sunglasses-face-stand-alone-men-s-1424065/

terms, a design has decorum if it is appropriate for its social context.[9] For example, a king's residence should look palatial because it is a royal palace, a school should look sober and solemn because those are seminal qualities of learning, and a home should look comfortable and secure because it exists to shelter a family unit.

In rough terms, a design has decorum to the extent that it:

1. Appears appropriate for its kind,
2. Belongs in its setting, and
3. Reflects its user's place in the social order.

Although the term *decorum* arises mainly in architecture, the concept may be applied to any design matter where appearances are important.

---

[9]Johnson (1994), pp. 226–229.

Case Study: Pixar HQ

Consider the Pixar headquarters building near San Francisco. The owner, Steve Jobs, wanted a modernist glass-and-steel design. However, Pixar co-founders John Lasseter and Ed Catmull felt that such a building would convey the wrong message about the company[10]:

> Lasseter and Catmull also resisted the idea of a minimalist, glass-and-steel headquarters. It didn't fit with either their industrial neighborhood or the rich, colorful, fantastical work being done by Pixar employees. "Pixar is warmer than Apple or NeXT," says Lasseter. "We're not about the technology, we're about the stories and the characters and the human warmth." They voiced their concerns to Tom Carlisle and Craig Paine, the architects Steve had hired for the job. Carlisle and Paine hired a photographer to shoot the brickwork of the lofts in the surrounding neighborhood, and in San Francisco. Then, at the end of one of the days when Steve was working from Pixar's Point Richmond headquarters, they laid dozens of those photos out on the table of a conference room. "He walked in and I remember him looking at all these beautiful photographs, all the details, and he walked around and around," remembers Lasseter. "Then he looked at me and he goes, 'I get it, you guys are right. John, you're right.' He got it, and he became a giant advocate for that look."

Jobs realized that a steel and glass structure would be out of place on this site. It would stick out among the red brick buildings in its vicinity. Jobs went to the extent of obtaining bricks from a foundry on Arkansas that used a palette similar to that found on the neighbouring structures.

The building also displays decorum in other ways. It makes use of wooden structural members and (polished) I-beams that echo the architecture of the factory that used to stand on the site.

Yet, it is different than its neighbors in ways that also reflect decorum. For example, whereas nearby structures are built right on the sidewalk, the Pixar headquarters is set on a large lawn. This difference reflects the greater size and fame of Pixar compared to the smaller companies that occupy nearby sites.[11]

Case Study: Monster Home

The term *monster home* is often applied to a house that fails to display decorum, usually by being too large for its surroundings. Consider the Elbasiouni house in Brampton, Ontario (Fig. 5).[12] In 2012, Ahmed Elbasiouni received a permit and began construction of a new house in a suburban neighborhood. In February 2013, neighbors complained to city authorities that the building, at three stories and 6600 sq. ft., was too large for the area, which consisted mostly of bungalows or two

---

[10]Schlender and Tetzeli (2015), pp. 331–332.
[11]Cf. Office Snaphots (2012).
[12]Grewal (2015).

Fig. 5 Brampton "Big blue house", in legal limbo. Courtesy of Fazal Kahn/Brampton Focus. URL: https://www.youtube.com/watch?v=BP-3Pm6Tjxw

story houses in the 1400–2000 sq. ft. size range. The City of Brampton revoked the building permit, arguing that it had been issued in error. Construction stopped and the half-built structure remains in limbo as Mr. Elbasiouni and the City contest the matter in court.[13]

There a number of ways in which this house appears to lack decorum. First, it is considerably larger than the other residences around it, and the one it replaced, being more the size of a multiplex than a single-family dwelling. Neighbors complain that it "towers" over their residences. Its footprint also occupies most of the property, unlike the neighboring houses, which sit behind small lawns.

Also, a representative of the neighborhood group MississaugaWatch describes the Elbasiouni house as "contemptuous" and "pretentious." In other words, some people infer from the scale of the building that the owner has inflated ideas about his status within community and resent him and his house for it.[14]

Q: What other designs exhibit decorum? Do not exhibit it?

[13]Grewal (2016).
[14]MississaugaWatch (2013).

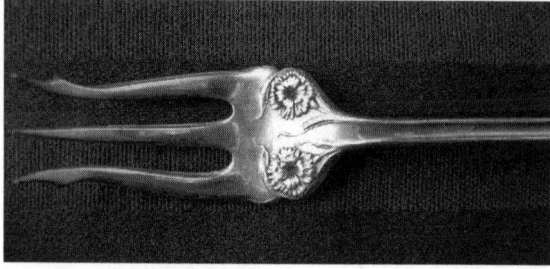

Fig. 6 Antique 1908 silverplated pickle fork. Photo by Karen/Wikimedia commons. URL: https://commons.wikimedia.org/wiki/File:Antique_1908_Silverplate_Pickle_Fork_from_W._R._Keystone.jpg

From a structuralist perspective, a design exhibits good style when it has decorum and exhibits bad style otherwise. However, there are potential problems that may arise with attention to decorum. For example, decorum places much emphasis on conforming to traditional or established practices. Sticking to the old ways may make acceptance of a design easier but it may also tend to stifle innovation.

Also, attention to decorum can become fussy or mannered. For example, by the late nineteenth century, forks had diversified into many different forms, each regarded as proper for eating a particular kind of food. There were oyster forks, lobster forks, salad forks, terrapin forks, berry forks, lettuce forks, sardine forks, pickle forks, fish forks, and pastry forks, among others (Fig. 6).[15] The point of having so many forks was for well-to-do people, who could afford dozens of kinds of forks in their cutlery services, to show off how sophisticated they were. However, the fussiness of having so many different kinds of forks became overwhelming and the number of fork types has since dropped to just a handful.

Commercialism: Style is Marketing

We have examined views that style should reflect strictly functional goals or that it should be expanded to include users' social goals. Another set of goals that designs often respond to are commercial goals. That is, the appearance of a design often reflects an effort to attract buyers and increase market share for its producers.[16] On this view, good design means success in the marketplace.

Although many factors influence how people respond to designs as commercial products, one important element is known as *branding*. In current parlance, branding refers to how consumers associate the form of a design with its producers and, indirectly, with other consumers.

---

[15]Goldsmith (2012).
[16]Bloch (1995).

Of course, producers normally want consumers to like the style of their products and thus have a positive impression of the producers. For this purpose, all that is required of products is that they appear pleasant.

However, producers also normally want consumers' impression to be clear. That is, they want the positive impressions of consumers to apply to them but not to their competitors. In that way, producers can reap more of the reward that comes with making a good impression.

To accomplish this marketing goal, product style should satisfy the following criteria:

1. *Identifiability*: The appearance of a design allows viewers to identify it with its source; and
2. *Differentiability*: The appearance of a design allows viewers to distinguish it from designs from other sources.

In satisfying these criteria, a design helps potential buyers to identify it with its producer and to avoid confusing it with designs from other producers.

Case Study: Coca Cola Bottles

Most people in the developed world would instantly recognize a Coca Cola bottle. This situation is no accident. The bottle itself is designed to evoke this response.

The Coca-cola bottle provides a good example of branding. For identifiability, the bottle has a curvy shape and fluted exterior, interrupted by a logo in on a smooth surface written in an old-fashioned hand. People have learned to associate these characteristics with a particular type of beverage.

For differentiability, other containers of similar beverages are at least somewhat different in form, either less curvy (more cylindrical, perhaps), and colored and inscribed differently, e.g., a Pepsi bottle.

The modern style of the Coca Cola bottle was introduced in 1916 precisely to achieve these effects. Before that time, the Coca Cola bottle was little more than a glass cylinder with a small spout, which resembled the bottles of many other soda pop producers. The president of the company commissioned a new bottle design in order to help the company's product to stand out[17]:

> We need a new bottle—a distinctive package that will help us fight substitution…we need a bottle which a person will recognize as a Coca Cola bottle even if he feels it in the dark. The bottle should be shaped that, even if broken, a person could tell what it was.

The shape of the new bottle was modeled on the profile of a popular woman's skirt. It is also distinguished by vertical fluting, which is interrupted by a band featuring the company name in raised letters.

---

[17]Lidwell and Mancasa (2009), p. 48.

Branding of this sort may be especially important in bulk goods such as soft drinks, all of which would look alike on a store shelf if it were not for differences is packaging. Also, by helping people to distinguish their product from others, e.g., Pepsi, branding enables Coca Cola to compete with rivals on grounds other than price. That is, Coca Cola can charge consumers more money for their products than rival producers can charge because loyalty to the brand makes consumers less sensitive to price.

> Q: What other products have strong design brands?

Exactly how branding works and how valuable it is are matters of dispute.[18] However, there is little doubt that many producers invest heavily in branding their products through their form.

Fashion

The logic of branding also leads to another marketing practice that can be labeled as *fashion*. In this sense, fashion refers to the practice of imitating the design style of another brand. It is not unusual for producers of goods who seek to increase their market share to imitate the style of the goods of the market leader.

For example, Wishing Well was a brand of cola make by National Dry Ltd. of Canada. Its bottle bears many significant similarities to the Coca-Cola bottle. It is made of clear glass, sports a waistline at a similar height, fluting down the body, and band for the product name made in raised lettering. The main difference is that the fluting is twisted rather than straight. It is clearly an imitation of the more popular brand.

The strategy of imitation seems straightforward. The similarity of packaging shapes informs consumers that the content of the bottle is a type of cola, similar to that of Coca-Cola. Given that imitators typically sell at a lower price point, consumers may decide that they would be happy to drink the imitation and keep the price difference in their pockets.

This "knock-off" approach to style could be considered an instance of free riding. That is, the imitator hopes to cash in on the hard work that the market leader has done to establish a positive impression on consumers. Coca-Cola put much money and effort into establishing a good reputation with consumers. By imitating their product design, Wishing Well hoped to put some of that good reputation to work for them.

---

[18]The Economist (2014).

Q: What other example of "knock-offs" can you think of? Are they successful?

Critique: Can Style Without Function Be Good?

From a modernist perspective, it is an error to manipulate appearances in design beyond the needs of functionality. In *Omit the unimportant*, Dieter Rams repudiates considerations of both social signaling and marketing. Trendsetters and marketers set up criteria for good design that are unrelated to functionality yet tend to interfere with it. Such matters are distracting for designers and confusing for users, he argues.

Changing tastes in eyeglasses provide an interesting example. In the 1990s, fashionable eyeglasses tended to be round, a design revived from earlier times. Harry Potter wore frames of this type. In the 2000s, the trend has gone to rectangular frames. This change in frame style seems to be due to changes in taste or the desire to stimulate sales and not because rectangular frames are better at correcting vision problems than round ones. Surely, this practice is simply manipulative and wasteful!

Jason Potts argues that the use of style to achieve social goals may be justified, notwithstanding objections like those made by Dieter Rams.[19] He argues that social goals play a beneficial role in consumers' selection of designs. In early models of consumer behavior, economists assumed that consumers cared only about the utility that products had for them alone. If Karl buys a Coca-Cola, for example, then that is because he thinks Coca-Cola will satisfy his preferences more than any other beverage available to him.

Potts points out that this assumption is false. Consumers care about not only what they think about a given product but also about what other people think of it. For example, Karl may buy a Coca-Cola not so much because he likes it more than the alternatives. Instead, Karl's friends are Coca-Cola drinkers and he wants to impress them. This perspective gibes with our earlier discussion of the role of social psychology in good design.

In addition, though, Potts argues that seemingly arbitrary changes in design resulting from social pressures are not simply manipulative and wasteful but are socially beneficial, a position he calls *fashionomics*.

To make this case, Potts compares changes in stylistic taste to economic recessions. Economic recessions, although they can be painful, can have long-term benefits for productivity in an economy. In economic good times, capital tends to

---

[19]Potts (2007).

get tied up according to the status quo. The profitability of established businesses leads people to prefer investments in those businesses to riskier ventures. The effect is to diminish innovation because that would upset or disrupt the status quo.

During an economic downturn, the status quo looks less attractive. As a result, people liquidate their assets and make new bets by investing in new ventures. Of course, many such ventures fail. Yet, many succeed, fostering greater innovation. In short, a recession can re-organize an economy in a way that is more productive than before. The result is an overall win for that economy.

By analogy, Potts argues that this logic about capital investment applies also to consumption. At a time when a style trend is entrenched, consumers tend to buy products that conform to that style. For example, when round glasses are popular, people tend to stick with those. The popularity of a style tends to suppress innovation in other stylistic and also functional possibilities.

When tastes change, the result is like an economic recession. Consumers get rid of their now unfashionable goods and invest in new ones, looking for the next "in" style. For example, as the consensus that eyeglasses should be round fades, people throw out or sell their round glasses and buy ones featuring alternative shapes. The result is greater experimentation with new designs, allowing consumers to find innovative products that work better for them. The result is an overall win for society.

In brief, changes in style can be advantageous in the following ways. First, consumers who earlier adopted goods that were not popular get a new chance to buy their way up the social ladder. For example, people who did not buy round eyeglasses in the 1990s can buy rectangular ones in the 2000s, thus boosting their own social standing. That is a win for them.

Second, as consumers become more open to trying new things, producers have a better chance of getting them to try out innovative new products. That is a win overall where those innovations make people more productive and their producers more wealthy.

Of course, any change in style can cause losses for some consumers. First, consumers who used to have stylish goods may invest in new goods that prove unpopular. For example, consumers wearing round glasses in the 1990s may have bought horn-rim glasses in the 2000s, thinking that horn-rim glasses would be the next big thing. Since that turned out not to be the case, the social standing of those consumers would actually drop.

Second, producers who made designs that consumers used to like may find that consumers reject their new offerings. For example, BlackBerry once made smartphones that were popular, featuring small, ergonomic keyboards. However, once the trend to smartphones with glass fronts became established, BlackBerry was unable to follow the trend. Even though they developed new and functional smartphones, they lost market share to more trendy competitors.[20]

---

[20]McNish and Silcoff (2015).

So, every turnover in style results in wins and losses. For Potts's argument to succeed, the wins from changes in style must outweigh the losses.

Q: Is Potts correct? Or, would we better off without socially-caused style turnovers?

References

Bilton, N. (2013, April 23). *The flattening of design.* Retrieved April 24, 2013, from The New York Times: http://bits.blogs.nytimes.com/2013/04/23/the-flattening-of-design/?_r=1

Bloch, P. H. (1995). Seeking the ideal form: product design and consumer response. *Journal of Marketing, 59*(3), 16–29.

Byrnes, M. (2013, July 8). *The many disguises of California's cell phone towers.* Retrieved July 20, 2013, from The Atlantic: Citylab: http://www.citylab.com/design/2013/07/many-disguises-californias-cell-phone-towers/6124/

Chan, C.-S. (2015). *Style and creativity in design.* Berlin: Springer.

Conway, H., & Roenisch, R. (2006). *Understanding architecture: An introduction to architecture and architectural history* (2nd Edition ed.). London: Routledge.

Gardiner, E. N. (1930). *Athletics of the ancient world.* London: Oxford University Press.

Goldsmith, S. (2012, June 20). *The rise of the fork.* Retrieved May 10, 2016, from Slate.com: http://www.slate.com/articles/arts/design/2012/06/the_history_of_the_fork_when_we_started_using_forks_and_how_their_design_changed_over_time_.html

Grewal, S. (2016, April 7). *Brampton monster home owner suing city for $20M.* Retrieved May 9, 2016, from Thestar.com: https://www.thestar.com/news/gta/2016/04/07/brampton-monster-home-owner-suing-city-for-20m.html

Grewal, S. (2015, Jan 8). *Unfinished Brampton 'monster' home remains in limbo.* Retrieved Jan 10, 2015, from Thestar.com: https://www.thestar.com/news/gta/2015/01/08/unfinished_brampton_monster_home_remains_in_limbo.html

Johnson, P.-A. (1994). *The theory of architecture: concepts, themes and practices.* New York: John Wiley.

Lidwell, W., & Mancasa, G. (2009). *Deconstructing product design.* Beverly, MA: Rockport Publishers.

Lovell, S. (2011). *Dieter Rams: As little design as possible.* London: Phaïdon.

McNish, J., & Silcoff, S. (2015). *Losing the signal: The spectacular rise and fall of blackberry.* Toronto: Harper Collins.

MississaugaWatch. (2013, Feb 14). *Brampton, Ontario MONSTER Monster home really *is* a MONSTROSITY.* Retrieved May 12, 2013, from Youtube.com: https://www.youtube.com/watch?v=DHg-EFnNIIM

Office Snaphots. (2012, July 16). *Pixar Headquarters and the legacy of Steve Jobs.* Retrieved July 8, 2015, from Office Snapshots: http://officesnapshots.com/2012/07/16/pixar-headquarters-and-the-legacy-of-steve-jobs/

Potts, J. (2007). *Fashionomics. Policy magazine, 23*(4), 10–15.

Rams, D. (1984). Omit the unimportant. In V. Margolin (Ed.), *Design discourse: history, theory, criticism.* Chicago: University of Chicago Press.

Schlender, B., & Tetzeli, R. (2015). *Becoming Steve Jobs: The evolution of a reckless upstart into a visionary leader*. New York: Crown Business.

The Economist. (2014, Aug 30). *What are brands for?* Retrieved Aug 30, 2014, from Economist.com: http://www.economist.com/news/business/21614150-brands-are-most-valuable-assets-many-companies-possess-no-one-agrees-how-much-they

Culture

Abstract Besides expertise in social psychology, expertise in culture may contribute towards the aim of good design as rational design. Culture involves the kinds of expectations that people in a given social group have about appropriate thinking and behavior. There are at least three ways that knowledge of culture might relate to good design. The first is that good design is universal, implying that culture is of little relevance to good design. This view is associated with modernism, on which a single, modern and industrial way of life was viewed as a universal good. Thus, good design should reflect this lifestyle. The second relation is that good design is contextual, that is, that good design means adapting designs to the various expectations that people in different cultures may have. An obvious example would be a software agent that users can address in their own language, rather than having to interact in a single language, such as English. The third relation is that good design accommodates innovation, at least within limits. Industrial designer Raymond Loewy recommended this view. He argued that good designs challenge people's expectations but without going so far as to alienate them from those designs. He called this idea the MAYA principle.

Social Groups

In the previous chapters, we have explored some social aspects of assessment of good design. People's experience with designs varies depending on how they think others will view them. Also, they often expect to communicate to other people through the designs that they use. These are aspects of human psychology that must be borne in mind when assessing good design.

One fact that becomes apparent in that exploration is that individuals are not all the same in their social views or responses. Neither are they all different. In other words, people can often be thought of as belonging to a variety of social groups. Membership in a social group can, in turn, be used to understand and anticipate how people will respond to a given design.

© Springer International Publishing AG 2017
C. Shelley, *Design and Society: Social Issues in Technological Design*,
Studies in Applied Philosophy, Epistemology and Rational Ethics 36,
DOI 10.1007/978-3-319-52515-0_5

Consider again the example of the Bugaboo Frog stroller. One reason to consider the stroller socially technotonic is because users can see themselves as members of a group that includes Gwyneth Paltrow, Gwen Stefani, Hugh Jackman, or Kate Middleton. Which social groups people belong to is important to their sense of identity and important for them to display through their possessions.

Busch also notes how different groups respond differently to stroller designs:[1]

> The Japanese prefer lightweight strollers, while those sold in England are rarely equipped with sunshades. Strollers used in France don't have rain shields, which are thought to be overprotective.

All the differences mentioned above relate to culture. That is, they describe preferences that people have in stroller design in terms of the cultural groups to which they belong. So, analogous to the case of social psychology, assessment of good design depends upon knowledge of culture. The purpose of this chapter is to characterize what culture is and different views on how it relates to good design.

Culture

To understand the importance of culture for assessment of good design, we need a reasonable definition of what a culture is. Here is a definition of culture from the literature on design of instructional materials[2]:

> [Culture is] the patterns of behavior and thinking by which members of groups recognize and interact with one another. These patterns are shaped by a group's values, norms, traditions, beliefs, and artifacts.

In other words, culture consists in membership in groups where each group is distinguished by shared beliefs, practices, and so on.

This definition of culture is quite broad. Nearly any social group to which people belong might be said to have a culture attached to it. A nation would be a good example. Consider Canadian culture, for example. Canadian culture might be characterized according to the definition in the following terms:

1. Canadian values include diversity and peaceful coexistence. This is not to say that Canadians always achieve these values, just that they are recognized as national ideals by Canadians on the whole.
2. A famous Canadian norm is polite behavior, centered on the word *sorry*, for which Canadians are notorious. In the comedy movie *Canadian Bacon*,[3] American commandos burst through a crowd of Canadians in Toronto. As each Canadian is pushed violently aside, they apologize for being in the way.

---

[1]Busch (2004), p. 32.
[2]Scheel and Branch (1993), p. 7.
[3]Canadian Bacon (1995).

Fig. 1 The doughnut, a beloved Canadian artifact. Photo by Evan Amos/Wikimedia commons. URL: https://commons. wikimedia.org/wiki/Category: Doughnuts#/media/File: Dunkin-Donuts-Chocolate-Sprinkled.jpg

3. As with many nations, Canadians celebrate the nation's birthday—July 1—with fireworks.
4. Canadians believe that Canada produces the world's best ice hockey players. The fact that those players do not always win world championships does not diminish this belief.
5. Some obvious Canadian artifacts include the toque, the beaver and the maple leaf. Also, Canadians are extremely fond of doughnuts, eating more of them, and having more doughnut stores per capita, than any other nation (Fig. 1).[4]

Any social group could be subject to the same sort of analysis.

Q: Use this definition to describe the culture of other social groups, especially some that you belong to.

Culture and Good Design

Historically, designers and design scholars have taken different views on the role and importance of culture in good design. In this chapter, we examine three approaches concerning the relation of culture and good design:

---

[4]Harrington (1994).

1. *Modernism*: Culture is inessential to good design.
2. *Contextualism*: Fit of designs with cultural expectations is important.
3. *Progressivism*: Designs should challenge the cultural status quo—somewhat.

The strengths and limitations of each perspective are considered.

Modernism

One response to the challenges posed for designers by culture is to reject culture as an essential consideration. That is, good design ignores the vagaries of culture as much as possible. How can this be?

The most important school of design to follow this route is known as *modernism*. Probably the most influential early modernist was Le Corbusier (1887–1965) (Fig. 2). He was an architect who was most active in the early and middle parts of the 20th Century and greatly influenced architecture and urban design in Europe and North America in the 1930s and later.

Le Corbusier argued that all branches of design, including architecture and urban design, should imitate the minimalism of modern machinery.[5] In his view, the kind of design found in industrial machinery like grain silos or mass-produced goods like cars was the only sort of design appropriate for the modern world. Furthermore, he

Fig. 2 Le Corbusier (1887–1965), born Charles-Édouard Jeanneret-Gris, was trained in engraving but became a most influential architect and urban designer in the early modernist movement. Photo by Holger Ellgaard. Detail of URL: https://commons.wikimedia.org/wiki/File:Lallerstedt_Corbusier_Tengbom_1933.jpg

---

[5]Corbusier (1927).

felt that the nature and requirements of modern life are the same everywhere. No matter where anyone lives, modern living for the masses means living in large apartment blocks, working in lofty office towers, and commuting to and fro in cars driven on superhighways.[6] The job of modern design was to embody this style of living, which it would help—or compel—people to adopt.

The epitome of this view was found in Le Corbusier's proposed *Plan Voisin*, a model of how Paris could be renovated along modernist lines. In this plan, staged for an exhibition sponsored by the French car and airplane manufacturer Voisin in 1925, Le Corbusier proposed to level much of Paris and replace it with a grid of sixty-story towers linked together by a network of highways and airports.

These claims have at least two implications for the role of culture in good design. First, if modern living is the same for everyone, then there is no need to adapt designs to reflect cultural differences. Instead, the point of good design is to assimilate people with different lifestyles to a single and unique conception of modern life. Attention to cultural variation in design would be retrograde.

Second, good design is universal and standard instead. Modernist designers tend to prefer modern materials, such as concrete, steel and glass, over local materials such as timber, stone, or rammed earth. Of course, deployment of modern materials must vary with function and the demands of the local, physical environment of a structure. Considerations such as local history and tradition are viewed as being of secondary importance, at best.

Case Study: La Villa Savoye

In order to promote industrialized living, Le Corbusier compared the design of buildings to that of industrialized or mass-produced goods. For example, he was fond of the claim that a house "is a machine for living." In other words, even a house should be designed as if it were something that would roll off a conveyor belt in a (very large) factory.

It was easy to apply this view to large apartment buildings where every unit is much the same as the next and is built in large numbers. However, Le Corbusier extended this scheme even to special architectural commissions, such as La Villa Savoye (Fig. 3). This house was a unique work commissioned by the wealthy Savoye family of Paris and was never meant to be repeated on an industrial scale. Even so, Le Corbusier designed it as a machine for living.

Its spaces are each assigned a basic function of modern living for a wealthy family. The garage and servants' quarters lie on the first floor. This arrangement is functional because cars and servants require direct access to the outside world to do their jobs. On the next floor are the kitchen, bathrooms, living room and sun room. This floor is larger than the first floor because the family required more space for its

[6]Meikle (1979), pp. 29–32.

functions than did the servants. Rather than expanding the walls of the first floor needlessly to support the second, the extra space is supported by small, external columns called pilotis. On the roof is the recreation area and garden.

The materials are all concrete, steel and glass. Thus, the house looks nothing like a traditional French dwelling. In fact, Le Corbusier used the design of passenger compartments on cruise ships, such as the HMS Aquitania, as a model (Fig. 4). The strip windows, support poles, and the funnel-like wall on the roof all echo elements of cruise ship design. This model is appropriate because cruise ships, like tall apartment buildings, provide accommodations in large quantities.

Q: What other designs are modernist in this sense? Are they good designs?

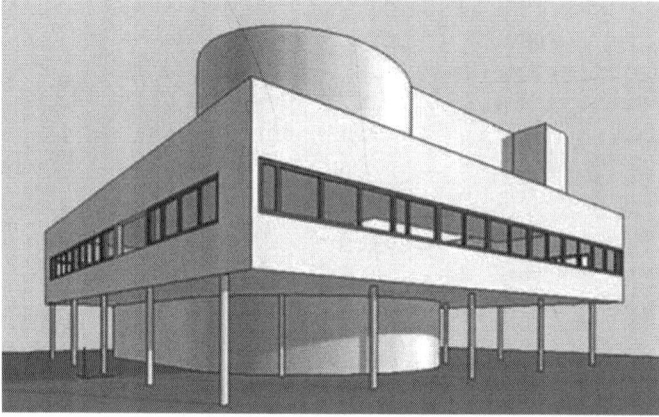

Fig. 3 Sketch of La Villa Savoye. Photo from Cameron Shelley: Models and Ideology in Design, In: L. Magnani, T. Bertolotti: Springer Handbook of Model-Based Science (Springer, Cham 2017)

Fig. 4 RMS *Aquitania*, Cunard Line. Photo from the United States Library of Congress. Detail of URL: https://commons.wikimedia.org/wiki/File:SS_Aquitania.jpg

Contextualism: Fit with Cultural Context

Le Corbusier's view demands that people conform to a universal ideal of modern living. An important implication of this view is that good design minimizes or eliminates cultural differences in order for the ideal to be achieved. Support for Le Corbusier's view can perhaps be found in the widespread attraction that modern, Western material culture has exerted around the world. One hundred years ago, automobiles, skyscrapers, office blocks and other trappings of modern life were restricted to a small number of cities in Europe and North America. Today, they are nearly universal.

In spite of this fact, there are reasons to doubt Le Corbusier's vision. Modern nations themselves contain diverse social groups that maintain distinctive cultural traditions. Also, different modern nations maintain different preferences in design. For example, consider attempts by American manufacturer Kohler to encourage Americans to adopt high-tech, Japanese toilets such as the Numi. This toilet has sensors that open and close its lid automatically when users approach or depart, programmable foot and seat warmers, an iPhone docking station, and a remote control. In addition it has a bidet function with a blow dryer for personal hygiene. Such features are common on upscale Japanese toilets; Japanese washrooms are often unheated, making a seat warmer highly desirable.[7] However, the toilet was ridiculed as odd, foreign, and over-the-top by the American late-night talk-show host Conan O'Brien.[8] The concept of a toilet featuring non-plumbing related functions does not exist in American culture and so appears laughably out of place.

Case Study: Insect Flour

An interesting example of cultural adaptation comes from a company called the Aspire Food Group. The group was founded by five MBA students at McGill University in Montreal. The group set as its goal to find a way to help people in developing nations to maintain a sustainable food supply. Of course, this problem is a huge one, so they pursued the more modest goal of providing a small kit that can provide a cheap and steady source of food protein.

Their product is a system for raising insects to be milled into nutritious flour. Insect farmers can buy a farming container and a starter population of insects. The Group won a $1 m grant from the Clinton foundation, and planned to produce 10 m tons of grasshopper flour in Mexico by the end of March 2014.[9]

Nutritionally, the plan makes sense. Insects provide crucial dietary proteins, and can be raised easily and in quantity by small farmers.

---

[7]See George (2008).

[8]Stoiber (2011).

[9]McCausland (2015).

However, identifying the right insects required the students to comply with local preferences[10]:

> "Power flour," as it's called, is going to be made with only "locally appropriate bugs." In Mexico, that means grasshopper. In Ghana, it'll be the palm weevil; and in Botswana, caterpillar.

> Ashour says originally the group wanted to just stick with the cricket. "When this was in the armchair phase, we liked the idea of taking one insect and popularizing it everywhere." The cricket, he says, "has a great resume," in that it's easy to farm, and can be found almost anywhere. "But in Mexico, people don't eat crickets; they do eat grasshoppers."

The insect of choice depends on which bugs the locals are accustomed to eating, in spite of any disadvantages such diversity might bring to the production process.

Interestingly, the group is also attempting to adapt their cricket farming design to the American market. Naturally, a substantial obstacle to this endeavor is the fact that Americans do not have a tradition of eating insects, at least, not deliberately.

Case Study: Jaipur Foot

Another good example of contextual fit and good design is provided by the Jaipur foot (Fig. 5). The foot is a prosthetic limb made of rubber and wood designed by a sculptor, Ram Chandra, and an orthopedic surgeon, P.K. Sethi (who had hired him) in 1968. Chandra was hired to teach art to polio victims at a hospital in Jaipur. He noted that many of the victims, who were amputees, could not wear western prostheses offered by the hospital. Thus, they simply did without, which greatly limited their mobility.

Chandra was considering how to design a better prosthesis one day when he got a flat tire. Taking the tire to a garage to be repaired, he noted how the repair was made with vulcanized rubber. This observation gave him an idea, which he refined with Dr. Sethi. The prosthetic foot is made from local materials, rubber and wood, and functions simply by virtue of the physical properties and arrangement of its components. It was made by a simple casting process, which local craftsmen are able to undertake reliably. The foot has been a success.[11]

Dr. Sethi notes cultural reasons why the Jaipur foot succeeded where Western designs failed[12]:

> "Wearing shoes, which were integral to the Western designed limbs, was uncomfortable in our hot climate," Dr. Sethi said. "Our people walk barefoot or in well-ventilated footwear.

> "We are essentially a floor-sitting people, requiring a range of mobility in our feet and knees which is not needed in the chair-sitting culture of the West. We should not expect our people to change their lifestyle because of a design we were forcing on them."

---

[10]Bichell (2013).

[11]Lidwell and Mancasa (2009).

[12]Bernstein (2008).

Fig. 5 Jaipur foot. Photo by Erin Collins. URL: https://www.flickr.com/photos/erincollins/638359780/

Western designs made assumptions about lifestyle grounded in Western culture, where people wear shoes outside and normally sit in chairs. Since these assumptions were crucial and did not apply in Jaipur, the Western designs were unsuitable. The Jaipur foot is a good design insofar as it conforms to the cultural and economic realities of life in rural India.

Q: What other designs that fit with their cultural milieu? Do not fit?

challenge culture.

Progressivism: What About Innovation?

Modernism and contextualism provide a pair of polar opposites. On the Modernist view of Le Corbusier, good design is design that demands conformity to a universal, industrial mode of life. On the contextualist view, good design is design that conforms to the differing norms and expectations of the prevalent cultural group.

Diff. btw Contextualist and Modernist.

Fig. 6 Steve Jobs (1955–
2011), was a founder and
president of Apple and a
prominent supporter of good
design as being in advance of
cultural expectations. Photo
by Matthew Yohe/Wikimedia
commons. Detail of URL:
https://commons.wikimedia.
org/wiki/Steve_Jobs#/media/
File:Steve_Jobs_Headshot_
2010-CROP.jpg

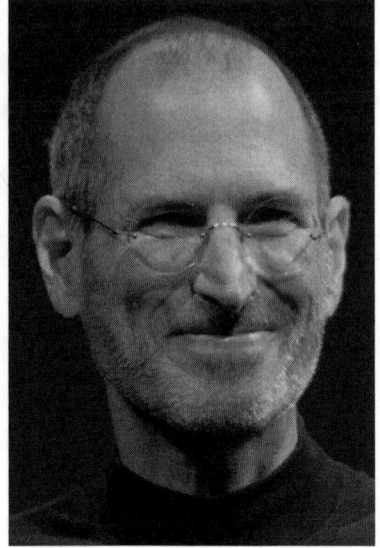

Although the Modernist view may be considered too rigid, the contextualist
view may be considered too flaccid. Some designers would argue that good design
does not mean simply mean giving people what they want or whatever simply
conforms to their expectations.

Consider the following anecdote about of Steve Jobs and his view of the value of
market research to discover people's expectations (Fig. 6)[13]:

> Some people say, "Give the customer what they want." But that's not my approach. Our job
> is to figure out what they're going to want before they do. I think Henry Ford once said, "If
> I'd asked customers what they wanted, they would have told me, 'A faster horse!'" People
> don't know what they want until you show it to them. That's why I never rely on market
> research. Our task is to read things that are not yet on the page.

Another way of putting this point is that good design sometimes requires
innovation, and some innovations will be surprising to people. So, the fact that a
design does not conform in every respect to existing cultural norms and expectations is not necessarily a bad thing. As Dieter Rams maintained, good design is
innovative.

As noted earlier, Rams did not clarify just when innovations are justified and
when they are not. However, the industrial designer Raymond Loewy thought that
he had the answer, namely that innovations are good design when they are novel
but not too strange.

---

[13]Isaacson (2011), p. 567.

Raymond Loewy and MAYA

Raymond Loewy (1893–1987) was an eminent industrial designer in the mid 20th Century (Fig. 7). Loewy was a Frenchman who served in the French Army Corps of Engineers in World War I. After the war, in 1919, he emigrated to the United States literally with only a few dollars in his pocket. However, he landed on his feet. In the U.S., he got odd jobs as an illustrator and a designer of store window displays.

Later, he got work designing consumer and industrial equipment, where he made his name. He worked on the design of office equipment, commercial logos such as the Lucky Strike cigarettes, and locomotives. However, his favorite subject was automotive styling, most notably including the Studebaker Avanti. He became something of a celebrity and his face even featured on the cover of Time Magazine.[14]

Loewy was the recipient of several honors and awards, including the Royal Designer for Industry award from the Royal Society of Arts in London (1937), one of the 100 "most influential Americans of the 20th Century" by Life Magazine (1972), and the Distinguished Achievement Award, American Society for Industrial Designers (1978), and was a founding member of the Industrial Designers Society of America (1946).

In his autobiography, Loewy tells the story of his professional development as an industrial designer.[15] In that book, he attempts to summarize some of the lessons

Fig. 7 Raymond Loewy (1893–1986), was born in France but migrated to the United States and became one of the pre-eminent industrial designers of the mid-20th century. Raymond LoewyTM/® by CMG Worldwide, Inc./www. RaymondLoewy.com

[14]Artzybasheff (1949).

[15]Loewy (1951).

that he learned. One central lesson he conveyed there was the MAYA principle: Most Advanced Yet Acceptable. Basically, this expression means that good design is innovative enough that it intrigues people but not so out-of-the-ordinary that it puts them off. Let us consider an example from Loewy's work.

Case Study: The Gestetner Mimeograph

In 1929, Loewy got his big break when he won a contract to redesign the Gestetner mimeograph, essentially an old sort of photocopier. Loewy had only three days develop his new design. Thus, he did not change the mechanism of the machine. Instead, he revised its chassis and "user interface."

The Gestetner mechanism was noisy and, because of the toner it used, was also smelly and apt to stain the clothing of its user (Fig. 8). For these reasons, it was regarded as a piece of industrial equipment and often relegated to utilitarian areas of office structures. Loewy designed a sleek cabinet for the Gestetner that made it look nicer and also muffled the noise it produced and protected the user from its toner and moving parts (Fig. 9). This case also simplified the appearance of the machine, making it feel more approachable. Users appreciated the enhanced usability of the new design and sales of the Gestetner improved as businesses bought multiple Gestetners for their offices instead of hiding a single unit in an out-of-the-way room.

In terms of the MAYA principle, the new Gestetner design was advanced in the sense that it brought the productivity of a piece of industrial equipment to an item of office equipment. This innovation was something that office managers readily appreciated. The redesign was acceptable in the sense that it presented office workers with an apparatus that appeared as tame as other office equipment they were used to, such as a file cabinet. Like a file cabinet, managers could place a new Gestetner in the office confident that their white-collar employees would have no objections and that visitors would be impressed with their presence.

Through his judicious balance between innovation and conformity in the redesign of the Gestetner, Loewy's design encouraged office managers to change the cultural status quo of the American office space.

Q: What other examples of MAYA can you think of?

Fig. 8 The Gestetner mimeograph, before it was redesigned by Raymond Loewy. Photo by Carsten Sadowski. URL: http://www.ebay.de/itm/Rare-duplicator-copy-machine-by-D-Gestetner-Tottenham-London-ca-1927-Model-15-/251955002475?nma=true&si=NNKp%252BhPnvzOiYsZyMB5kOEwaSYk%253D&orig_cvip=true&rt=nc&_trksid=p2047675.l2557

Culture and Advancement

On the MAYA principle, good design is not just giving people what they would find culturally acceptable. Instead, it is exploiting their sense of progress in order to get them to loosen their idea of what is culturally acceptable.

As the science-fiction author Bruce Sterling explains, Loewy's realization amounts to two claims.[16] First, people tend to want progress and innovation. Second, people tend to resist change.

---

[16]Sterling (2005).

Fig. 9 The Gestetner
mimeograph as redesigned by
Raymond Loewy. Photo by
Erwin Blok, courtesy of
stencilroloFlick/Flickr.com.
URL: https://flic.kr/p/
5NQPUk

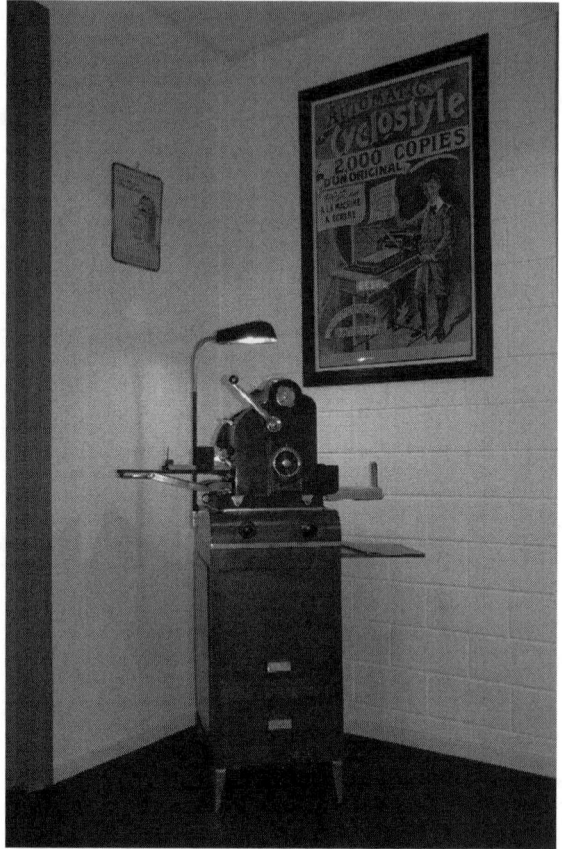

There is an obvious tension here that good design must navigate successfully.

As Sterling says, "Designers mine raw bits of tomorrow". In other words, they produce innovative technology, things that everyone in the future will have. People today want a piece of that future now, meaning they want innovation.

However, it is in the nature of culture that people hesitate to make drastic changes. Current production methods may not be compatible with highly futuristic technology, so industrialists will resist retooling their factories to make it. Many of the most influential investors will have stakes in the old technology. Potential customers may find new designs incomprehensible or even threatening.

Because culture makes many people resistant to change, they cannot be reasoned into adopting a new design. Instead, the design has to appeal to their aspirations for the future. As Sterling puts it, "the customers must be seduced". Designs should charm customers into adopting them, prompting them to imagine how much better life could be with the new design.

So, the art of MAYA is, as it were, taking things from the future, and presenting them in a way that fits with the present. To extend Sterling's metaphor, think of a designer with a time machine. The designer uses a machine to travel into the future and bring back future technology to the present day. According to the MAYA principle, the designer should go only a few years ahead. That way, the designs that are brought back from the future will still be recognizable to people today as advanced versions of present designs. If the designer goes too far into the future, then the designs brought back will seem weird and alien, a poor fit with current expectations and lifestyles.

Case Study: The Failure of the Airflow

An example of a design that failed to respect the MAYA principle was the Chrysler Airflow (Fig. 10).[17] It was introduced in 1934, a decade that brought great stylistic changes to car design, and not so much technical improvement. The innovative aspect of the Airflow was its aerodynamic shape. Especially compared to popular cars of the 1930s, such as the Ford V8 (Fig. 11), the Airflow looks as though it were designed in a wind tunnel. In fact, airplane design was widely admired and imitated in the design of cars, buildings, and furniture of the era. Airplanes had become noticeably more streamlined and efficient in the preceding years, and streamlining was widely viewed as the way of the future. So, it made sense to produce car like the Airflow that had many streamlined features including integrated and rounded wheel well covers, curved fenders and hood, tilted-back front windshield, and lower road clearance.

However, as Volti notes, the Airflow was unsuccessful and ceased production in 1937. One of the main reasons, according to Volti, was that the nose of the car had been brought far in front of the front wheels. Thus, the car had a "nose-heavy" look that was exaggerated by elimination of the grill in favor of a dramatically sloping front hood. Potential customers were turned off by the radical departure from automotive design norms of the time. The Airflow was just too advanced; people were not yet ready to fly their cars home.

The example of the Airflow illustrates that Lowey's MAYA principle enjoys some explanatory power. It can help to explain the failure of some products as well as the success of others.

Ahead of Its Time

Historically, it is interesting to note that many of the streamlined features of the Chrysler Airflow did appear in later cars. Curved bumpers, forward engine compartments, integrated headlights and wheel wells become commonplace on North

---

[17]Volti (2004), pp. 70–73.

Fig. 10 The streamlined
1936 Chrysler C-10 Imperial
Airflow. Photo by Greg
Gjerdingen. URL: https://flic.
kr/p/nALqRy

Fig. 11 The Ford V8, a more
conservatively styled
contemporary of the Airflow,
known in the 1930s for its use
by outlaws Bonnie & Clyde.
Detail of URL: https://flic.kr/
p/p23BpP

American cars by 1950. It might be said that the Airflow was simply too far ahead
of its time.

Consider the example of a coffee cup created by Italian designer Massimo
Vignelli (Fig. 12) for Heller Dinnerware. See Fig. 13. One innovative feature of the
cup was that the handle was an integral part of the cup itself rather than a separate
piece attached by glue. One consequence of this design was that the handle left a
notch in the rim of the cup. Vignelli relates that Heller received many complaints
about this feature because, when the cup was filled with coffee, the notch and handle
would allow hot coffee to spill through the notch and down the handle, which acted
like a gutter.[18] Owners complained about the resulting messes and scorched thumbs.

The problem was cultural in origin. Vignelli had designed the cup as a *demi-tasse*, that is, a cup that would normally be filled only part-way in his native Italy.
In the United States, at the time, it was normal to fill a *coffee cup* to the brim.
Vignelli and Heller reluctantly changed the design to fill in the notch and prevent
spills for American customers.

---

[18]Brew and Guerra (2012).

Fig. 12 Massimo Vignelli (1931–2014), was an Italian designer known for packaging, housewares, and furniture design, plus the wayfinding system for the New York City subway. Photo by Massimo Vignelli/Wikimedia commons. URL: https://commons.wikimedia.org/wiki/File:Massimo_Vignelli_2.jpg

Fig. 13 Outline drawing of demitasse designed by Massimo Vignelli for Heller Dinnerware. Note how the handle creates a trough at the lip of the cup. Drawing by Cameron Shelley

In retrospect, Vignelli denies that he made a mistake.[19] Instead, he explains that he was simply ahead of his time. In his view, filling a coffee cup to the brim is simply uncivilized and Americans should learn to treat coffee in the way that Italians do. Today, he continues, Americans have begun to do just that. "When you are ahead of your time," he comments, "then, by implication, many people are behind you."

Here, he explains his perspective in a radio interview:

---

[19]Newman (2013).

Interviewer: I'm wondering: how do you know when you're ahead of your time versus just being wrong?

Vignelli: I wasn't wrong. I wasn't wrong.

Interviewer: But people weren't able to drink their coffee in America for a while.

Vignelli: That is *good*. They do it once, then they learn, which at the end turns into an advantage. It's very rude to fill up the cup all the way to the top, you know, and so they learn how to be civilized by filling it up less.

> Q: Was Vignelli's design a mistake? What other designs have been "ahead of their time"?

Adapting Design: Appropriate Technology

Loewy's MAYA principle suggests a middle ground between the extremes of Modernism and contextualism. It suggests how good design may be design that neither ignores cultural differences in favor of universal ideals, nor rules out designs that are not merely what people expect or are comfortable with.

One further issue concerning good design and culture is how designs from one culture may be adapted to suit people in another. One form of this problem has been discussed in terms of *technology transfer*, that is, the adaptation of industrial designs from developed countries to meet the requirements of developing ones. For example, to enhance the productivity of the textile industry in a developing nation, western designers might install a factory with automated looms. An alternative approach would be to design more efficient hand looms of the type that people in the target country already operate.

Each approach has its advantages. The first would produce a high output and would pay off as long as textile prices remain buoyant. The second would employ more local workers and be understandable to and maintainable by them.

Around 1960, the British economist E. F. Schumacher advocated an approach called *appropriate technology* (AT). In his view, the best way to advance the economies of developing nations is not to deploy the latest designs from developed countries. Instead, smaller scale, intermediate technology should be applied. He argued that designs such as improved hand looms would be the better choice in the example above, since it would involve less training, require a smaller outlay, and take advantage of the labor pool that is often readily and cheaply available in developing economies.

There is no single definition of AT, but the following criteria are widely used to characterize this perspective on good design[20]:

1. Small scale;
2. Energy efficient;
3. Environmentally friendly;
4. Labor intensive;
5. Controlled by the local community;
6. Simple enough to be maintained with local expertise.

Designs with these qualities are good for developing nations because they are often short of reliable energy sources, good transportation, advanced technological expertise, and stable and supportive government structures. Thus, it is better to design items that require labor over expertise and can be controlled effectively at a local level rather than rely on large commercial or national institutions.

Case Study: The Nut Sheller

An instructive example of AT in practice is provided by the Malian (later Universal) Nut Sheller (Fig. 14). This device is a peanut-shelling machine designed by Jock Brandis to help Malian peanut growers to process their crop more efficiently and without causing environmental or governmental problems. In 2001, Brandis travelled to Mali to help repair a water treatment system. There, he noted the importance of peanuts to the local economy and promised to send an automatic peanut-shelling machine back upon his return to the United States. Finding that no satisfactory machine existed, Brandis designed one and returned with equipment needed to make copies of the design on site.

The sheller is a simple machine in which peanuts are fed into the top and then rolled between two inverted cones, the inner cone being rotated by a hand crank. The rolling action then separates the peanuts from their shells, which then drop out of the bottom into a basket. Winnowing allows the peanuts to be captured for later sale. One sheller can process about 50 kg of peanuts per hour.

The sheller is made from regular concrete set in a fiberglass mold and requires only some simple metal parts. It is robust and requires less than $50 of materials to make.

The story is told compellingly and in more detail in a video named *Peanuts*.[21]

Q: In what ways is the Sheller an example of AT?

---

[20]Hazeltine (1999), pp. 3–4.

[21]Harbury (2002).

Fig. 14 Rachel and the nut
sheller. Photo by Rex
Miller/Wikimedia commons.
URL: https://commons.
wikimedia.org/wiki/File:
RachelAndMachine.jpg

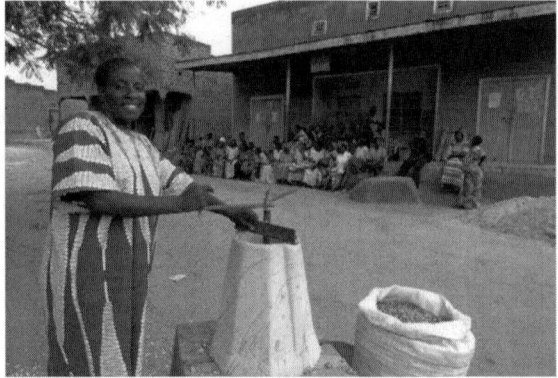

Besides these virtues, the design is also easily adaptable to other situations in which peanuts are grown as a cash crop. In addition, it helps in the achievement of beneficent social aims, namely the preservation of family units but also the emancipation of women from difficult and time-consuming manual work, freeing them to take more schooling.

Q: How does AT compare with the MAYA principle?

References

Artzybasheff, B. (1949, October 31). *Raymond Loewy: Oct. 31, 1949*. Retrieved September 28, 2015, from Time: http://content.time.com/time/covers/0,16641,19491031,00.html

Bernstein, A. (2008, January 8). *P.K. Sethi: Surgeon fashioned new limbs*. Retrieved September 28, 2014, from The Washington Post: http://www.washingtonpost.com/wp-dyn/content/article/2008/01/07/AR2008010703191.html

Bichell, R. E. (2013, September 27). *Students win seed money to make flour from insects*. Retrieved September 28, 2013, from NPR: http://www.npr.org/sections/thesalt/2013/09/27/226853288/students-win-seed-money-to-make-flour-from-insects

Brew, K., & Guerra, R. (Directors). (2012). *Design is one* [Motion Picture]. McNabb Connolly.

Busch, A. (2004). *The uncommon life of common objects: Essays on design and the everyday*. New York: Bellerophon.

Canadian Bacon (1995). [Motion Picture]. Peter Markle.

Corbusier, L. (1927). *Towards a new architecture*. (F. Etchells, Trans.) London: The Architectural Press.

George, R. (2008). *The big necessity: The unmentionable world of human waste and why it matters*. New York: Picador.

Harbury, M. (Director). (2002). *Peanuts* [Motion Picture].

Harrington, B. (1994, September 1). *The doughnut: Canada's unofficial sugary snack.* Retrieved September 23, 2012, from CBC news: http://www.cbc.ca/archives/entry/the-doughnut-unofficial-national-sugary-snack

Hazeltine, B. (1999). *Appropriate technology: tools, choices, and implications.* London: Academic Press.

Isaacson, W. (2011). *Steve Jobs.* New York: Simon and Schuster.

Lidwell, W., & Mancasa, G. (2009). Jaipur foot. In W. Lidwell & G. Mancasa (Eds.), *Deconstructing product design* (pp. 92–93). Beverly, MA: Rockport Publishers.

Loewy, R. (1951). *Never leave well enough alone.* New York: Simon and Schuster.

McCausland, P. (2015, September 24). *How to breed a tasty cricket.* Retrieved September 28, 2015, from The Atlantic: http://www.theatlantic.com/health/archive/2015/09/americas-cricket-farmers/406843/

Meikle, J. L. (1979). *Twentieth century limited: Industrial design in America, 1925–1939.* Philadelphia: Temple University Press.

Newman, B. F. (2013, October 11). *Massimo Vignelli on design and civility.* Retrieved September 28, 2015, from The Dinner Party download: http://www.dinnerpartydownload.org/design-is-one/

Scheel, N. P., & Branch, R. C. (1993). The role of conversation and culture in the systematic design of instruction. *Educational Technology, 33,* 7–18.

Sterling, B. (2005). *Shaping things.* Cambridge, MA: The MIT Press.

Stoiber, M. (2011, May 5). *The toilet that taught me an innovation lesson.* Retrieved January 18, 2016, from Huffington Post: http://www.huffingtonpost.com/marc-stoiber/the-toilet-that-taught-me_b_866379.html

Volti, R. (2004). *Cars and culture.* Baltimore: The Johns Hopkins University Press.

Social Contract

Abstract One of the limitations of rational design as a model of good design is that it omits moral considerations. Recall Dieter Rams's view that good design is aimed at making the world more humane. This criterion is not about how optimal a design is in the achievement of its function but rather about the quality of that function itself. The question is: Is the world that a design helps to bring about a good world? The concept of the social contract is relevant to addressing this question. A social contract is, typically, a body of rights that people observe so that they may thrive through collaboration and cooperation. Through cooperation, people may become better off than they would be by simply acting individually on their own behalf. On this view, the question of good design comes down to a matter of how well designs respect people's rights within some applicable social contract. Social contracts are often reflected in legal codes, such as safety regulations, that spell out how designs are expected to perform.

Rational and Moral Design

Recall that, up to this point, we have adopted Herbert Simon's ideal of good design as rational design. On this view, good design means design that provides an optimal solution to a given problem or, at least, as optimal a solution as knowledge permits.

However, this approach has some significant limitations. Recall Simon's view that design is not only a matter of means but also of ends. In that case, design evaluation involves not only assessing how goals are achieved but also assessing goals themselves. Since considerations of rationality apply only to means and not ends, evaluation of goals requires some new concepts.

Moral concepts are appropriate for evaluation of goals. Think of Dieter Rams's argument that good design is humane. This concept had little to do with how a design works and more to do with the world that a design is supposed to help create. In short, it is a moral concept. In this and subsequent chapters, we will explore the use of moral concepts for the job of design assessment.

C. Shelley, *Design and Society: Social Issues in Technological Design*,
Studies in Applied Philosophy, Epistemology and Rational Ethics 36,
DOI 10.1007/978-3-319-52515-0_6

Don't Be Evil

Consider the former corporate motto of Google, "Don't be evil," which it adopted in early 2000. This motto is clearly framed as an assessment of the design of Google's products, as well as the conduct of the company itself. It is also clearly not a rational characterization of good design, such as "Don't be ineffective" or "Don't be inefficient," or, more generally, "Don't be suboptimal." Instead, it invokes a moral concept. Here is how Paul Buchheit, Google employee 23 and originator of the motto, explains it (Fig. 1)[1]:

> It just sort of occurred to me that "Don't be evil" is kind of funny. It's also a bit of a jab at a lot of the other companies, especially our competitors, who at the time, in our opinion, were kind of exploiting the users to some extent. They were tricking them selling search results —which we considered a questionable thing to do because people didn't realize that they were ads.

In other words, other companies designed their search engines to mix advertising in with search results without making users aware of this fact. Although such a design might be effective in making money for Google, it was immoral, in Buchheit's view.

There are two points to note about Buchheit's concerns. First, his concern is aimed primarily at the goal of some search engine designers, that is, tricking users, and not at the means employed, that is, mixing ads in with search results incognito. Second, his concern is explicitly a moral one: He does not say that tricking users is ineffective, inefficient, etc. Instead, he argues that its aim is "questionable" or wrong. This evaluation, in turn, implies that the means employed in the search engine design is not morally acceptable either.

Interestingly, Google's parent company Alphabet, has replaced the old motto with the following[2]:

> Employees of Alphabet and its subsidiaries and controlled affiliates ("Alphabet") should do the right thing—follow the law, act honorably, and treat each other with respect.

The language is still largely moral in focus, including being right, lawful, honorable and respectful.

For the remainder of this book, we will examine concepts that help in framing moral evaluations of design, hopefully with a little more clarity than the motto "Don't be evil." In other words, we will be pursuing moral evaluation of designs from a technology-society perspective.

---

[1]Livingston (2007), p. 170.
[2]Alphabet (2015).

Fig. 1 Paul Buchheit,
Google employee 23 and
originator of their motto
"Don't be evil." Photo by
Robert Scoble/Wikimedia
commons. Detail of URL:
https://commons.wikimedia.
org/wiki/File:Paul_Buchheit.
jpg

Case study: The BroApp

The difference between a rational and a moral evaluation of a design is illustrated in
the case of the BroApp. The BroApp is a smartphone application designed for men
to send out automated daily text messages to their girlfriends in order to "maxi-
mize" the romantic quality of the relationship. Its creators describe it as a "clever
relationship wingman" that provides "seamless relationship outsourcing"[3]:

> The app includes other features to help increase the realism of its offerings:
>
> The developers also took steps to conceal the automation going on behind the scenes; in
> places designated "no bro zones," the app is automatically disabled. (After all, the jig is up
> if your girlfriend received an automatic text from you while you're at her place.) The app
> even has a rating system that lowers the risk of the same message being sent too frequently.

The creators, James and Tom, argue that since BroApp optimizes the efficiency
of romantic relationships, it is a good thing[4]:

> A guy starts using BroApp with his girlfriend, set to send a message around 12 pm each
> weekday. Guy observes that girlfriend is now much happier when he arrives home from
> work. Guy is no longer stressed about finding time during a busy day to text. Girl is much
> happier because her boyfriend is more engaged with their relationship.

---

[3]Selinger (2014).
[4]Selinger (2014).

On this view, everyone is more happy and no one is less happy. So, it sounds as though the BroApp is both rational, because it achieves its end in an optimal way, and moral, because it makes everyone happier than they would be otherwise.

Evan Selinger of the Rochester Institute of Technology argues that BroApp is a bad thing, the arguments above notwithstanding. Selinger argues that BroApp would undermine what makes relationships worthwhile in the first place[5]:

> Ultimately, the reason technologies like BroApp are problematic is that they're deceptive. They take situations where people make commitments to be honest and sincere, but treat those underlying moral values as irrelevant—or, worse, as obstacles to be overcome. If they weren't, BroApp's press document wouldn't contain cautions like: "Understandably, a girl who discovers their guy using BroApp won't be happy."

On his view, even if everyone is happier as a result of BroApp, its reliance on deception makes it immoral.

Q: Is BroApp a good design or a bad one?

Besides the concerns raised by Selinger, this case may also reflect a kind of sexism that pervades the technology industry. Why not have an app for women (BeauApp?) to use to deceive their boyfriends? If that design sounds unacceptable, then perhaps BroApp should be viewed in the same way.

Two Kinds of Good

The relationship between rational and moral evaluations of good design can be clarified by observing that the expression *good design* is ambiguous. In other words, the concept actually has (at least) two senses:

1. *Rational*: Good designs are ones that achieve their ends in excellent ways.
2. *Moral*: Good designs are ones that help to achieve excellent ends.

The first sense is the one that we have been exploring up until now. The second sense is the one that we are going to explore further.

In the BroApp example, the question, "Is BroApp a good design?" is answered in the first sense by noting how well it helps men to make their girlfriends feel that they are engaged in their relationship. For those women who see regular text messages from their boyfriends as evidence of engagement, the app may work well.

In the second sense, the goodness of the design is more debatable. Its designers argue that, since the app makes both parties happier, then it is morally acceptable. Yet, the use of trickery to deceive the girlfriends involved makes it morally

---

[5]Selinger (2014).

Fig. 2 A late 19th century Italian chimney sweep and his boy assistant. URL: https://commons.wikimedia.org/wiki/File:Bub_und_Meister.JPG#/media/File:Bub_und_Meister.JPG

problematic. Just as disguising advertising as search results would violate Google's motto, "Don't be evil," disguising canned text messages as live and spontaneous communications would be "evil" also.

Rational Versus Moral

To draw a line under the distinction between rational and moral evaluations of designs, it helps to see how designs can be good in one sense but not in another sense.

First, consider an example of a design that is rationally good but not morally good. In Victorian Europe, young boys were sometimes employed as chimney sweeps. This use of boys was rationally good because they could fit in the confined spaces presented by chimneys, they were light enough to climb high into flues (thus earning the name "climbing boys"), they could be obtained at very low prices, and their use was largely unregulated (Fig. 2).[6]

Morally, however, this treatment was not good. The boys were often abused by their masters, inadequately compensated for their work, and subject to fatal lung diseases from exposure to the dust and ash they breathed in during sweeping.

---

[6]Horn (1995).

Fig. 3 Winston Churchill, 31
December 1940, as Prime
Minister of the United
Kingdom. Churchill believed
that democracy is both the
best and worst form of
government. Photo courtesy
of the United States Library of
Congress. URL: https://
commons.wikimedia.org/
wiki/File:Winston_Churchill_
cph.3b12010.jpg#/media/File:
Winston_Churchill_cph.
3b12010.jpg

Put another way, the use of boys as chimney sweeps was optimal as a means of cleaning chimneys but was immoral due to the abuse of the boys involved.

Second, consider an example of a design that is rationally not good but morally good. Winston Churchill once made the following assessment of democracy as way to run a country[7]:

> Many forms of Government have been tried, and will be tried in this world of sin and woe. No one pretends that democracy is perfect or all-wise. Indeed it has been said that democracy is the worst form of Government except for all those other forms that have been tried from time to time.

This statement appears to be a contradiction. How can something be both the worst form of government but also better than the alternatives? Yet, Churchill was serious (Fig. 3).

Q: What might Churchill mean?

Given his experiences, it is easy to image how Churchill might have arrived at this view. In the years before World War Two, Hitler was able to turn Germany from a country sunk in deep economic depression into a military superpower. He accomplished this goal because, as dictator, he did not have to endure the distraction of people who disagreed with his policies.

---

[7]Churchill (2008), p. 574.

In contrast, Churchill had a difficult time persuading Britons to prepare for an armed conflict. He had to persuade people who denied that a war was imminent or who would prefer to put off Hitler rather than confront him. When war did start, Britain was far less prepared than Nazi Germany. A democratic political system can make it hard to achieve certain policy goals but it is more respectful of the rights of its constituents than is a dictatorship where the will of the leader is beyond question.

Together, these examples illustrate how rational and moral evaluations are different, and how it would be incorrect to confuse them.

Rights

Historically, the concept of a right has been important in moral evaluations. Simply put, a right is an entitlement that one person has for respectful treatment from others. In other words, having a right means that other people have a moral obligation to treat you in a certain way.

An influential account of rights was given by English philosopher John Locke (1632–1704) (Fig. 4). Locke wrote books on a number of topics and his views about government greatly influenced political reformers during the American and French Revolutions. According to Locke, all people enjoy three basic rights[8]:

1. *Life*: a right to personal safety;
2. *Liberty*: a right to non-interference from others;
3. *Property*: a right to exclusive access to resources.

People need these rights to thrive. A right to life seems obviously necessary, since people cannot thrive if others are entitled to kill them arbitrarily. A right to liberty is necessary too since people cannot thrive if others have a right to restrict them arbitrarily, e.g., by preventing them from finding food, shelter, or other necessities of life. Similarly, a right to property is necessary. A right to property is an entitlement to prevent others from using or taking a person's possessions. If others could arbitrarily confiscate anyone else's possessions—including the necessities of life—then no one could be assured of the resources needed to thrive.

If the rights to life, liberty and property sound familiar, it may be because the framers of the U.S. constitution had read Locke's work. The rights to life, liberty, and the pursuit of happiness mentioned in the Declaration of Independence were modeled on Locke's concepts.

---

[8]Locke (1689/1988).

Fig. 4 English philosopher John Locke, who promoted the concept of society regulated by the Social Contract. Photo courtesy of Wikimedia commons. URL: https://commons.wikimedia.org/wiki/File:Locke-John-LOC.jpg#/media/File:Locke-John-LOC.jpg

A Social Contract

Rights to life, liberty and property allow people to thrive in the sense that they are entitled to pursue their individual goals. However, Locke argued, people could thrive even better if they could cooperate and collaborate on common projects. In other words, there are advantages for everybody if people could organize their efforts and achieve things that they could not while acting solely as individuals. Certainly, technological achievements like the Apollo moon program were the result of individuals cooperating as a group and not merely going around pursuing their own agendas without regard to others.

In order to cooperate so fully, people would need a more elaborate package of rights to regulate their treatment of one another. Locke called such a package of rights a *social contract*. Like a business contract, a social contract is a deal that people make with each other to specify how they will treat each other and share things. It lays out what rights and responsibilities apply to everyone who is part of the deal. Such rights are typically specific to certain situations rather than as broad as rights to life, liberty, and property.

Case Study: Crosswalks

To see how a specific social contract can work, consider a crosswalk. Crosswalks come in different kinds, e.g., ones with signals versus ones with only pavement markings. However, all crosswalks embody a certain sort of social contract.

Fundamentally, a crosswalk identifies a piece of roadway that is to be shared by two groups of people, typically drivers and pedestrians. Each group wants access to the piece of road in order to cross it. The roadway could be shared on an everyone-for-themselves basis. However, this arrangement could be extraordinarily dangerous for pedestrians and, at least, nerve-wracking for drivers.

Instead, crosswalks are designed so that people access a roadway on a turn-taking basis. That is, each group takes turns using the roadway. At a signalized crosswalk, lights inform people when it is their turn to go and when it is their responsibility to stop. Sometimes, pedestrians must request a turn by pushing a button. The area of roadway subject to turn-taking is often painted with special markings.

In terms of a social contract, a crosswalk enforces turn-taking by specifying who gets the *right of way* and when. The right of way is a right that one party has to access the roadway while the other party is excluded. As such, it is a kind of liberty right: The party with the right of way has the liberty to enjoy the crosswalk without interference from the other party.

In other words, a crosswalk is designed to implement a social contract. It embodies and enforces a code of conduct that people observe in order to reap the benefits of access to a roadway through cooperative action.

> Q: What other designs enforce sharing by turn-taking?

Notice also that the right of way imposes a limitation on the rights of liberty of everyone seeking access to a stretch of roadway. That is, pedestrians enjoy the right of way only when drivers do not, and vice versa. Very often, a social contract creates conditions for cooperation to occur through compromise on people's basic rights. It is through such compromises that people can enjoy benefits that come with working together.

A Range of Social Contracts

Following Locke, people use the expression *the social contract* to refer to a basic set of rights that people obtain from being part of a state. However, the concept of social contract can be extended to cover any kind of arrangement that people have to regulate their conduct towards one another. These social contracts can be thought of as occupying a continuum depending on how deeply connected they are with the morality of our behavior (Fig. 5).

At the one end is *the* social contract that describes what constitutes basic, moral behavior, including rights to life, liberty, and property. After this is the set of laws and regulations that are spelled out by governments, courts, and other competent agencies. These laws, rulings, and regulations have moral force insofar as they help

Fig. 5 A continuum of social contracts, ranging from etiquette to the fundamental social contract of a society

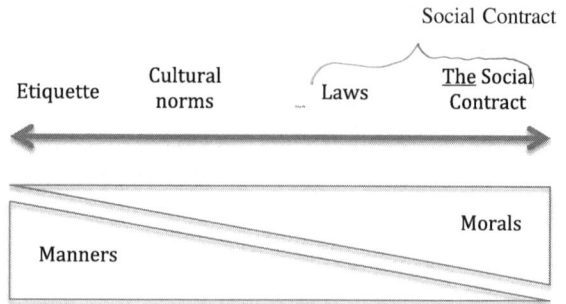

the government to realize its duty of establishing the enjoyment of basic rights for everyone in society.

After such institutional regulations come the social or cultural norms that exist within any society but that are not embodied in legislation. The moral force of cultural norms is less clear but they may carry some weight if they embody the collective and time-honored wisdom of a society. In some cases, however, cultural norms may not be of much moral significance.

Finally, there are matters of etiquette, that is, what is considered to be good manners. Good manners, such as which fork to use to eat dessert, often have little or no moral importance.

As such, social contracts about etiquette and cultural norms tend to fall under the concepts of culture, style, and social psychology that we have discussed earlier. The following discussion, then, focuses on social contracts that fall on the right-hand side of the scale above.

Q: Where do crosswalks fit on this continuum?

Case Study: Parking

Having looked at crosswalks, we can expand our repertoire of social contracts by looking at another contested and road-related resource, parking spaces. In most cities, public parking spaces are provided by the city for any driver who wishes to use them. That is, any driver who finds an available space, and who has enough money for any meter that may be present, has the right to park in the spot. Because parking spaces are in limited supply, drivers collectively spend a fair amount of time cruising city streets looking for parking that is available and convenient.

This way of allocating parking is on a first-come, first-served basis in the sense that the first driver to find an available space has the right to occupy it. In such a system, every driver has an equal right to each space; priority is assigned by time of arrival and no driver is able to pre-empt others.

A challenge to this social contract on parking has been put forth by an app service called MonkeyParking.[9] This service permits users to auction off their parking spaces. When a user is about to leave a parking space, they broadcast this fact on the service, which sends alerts to other users in the vicinity. Those users can then bid between $5 and $20 in order to be the next driver to park in the space. When the auction is won, the parker waits for the winner to arrive to take up the space. Given the scarcity of parking spaces in some areas, it is easy to imagine that drivers would be willing to pay for prime parking spots.

Allocating parking spaces by auction assigns priority to drivers not on an equal basis but based on their ability to pay. Drivers who might be the first on the scene of an available spot could be pre-empted by other drivers who outbid them.

Proponents of this approach argue that auctions are economically efficient, that is, drivers who most want the resource can get it whereas others who are not so desperate can get other parking spaces for a price they are willing to pay. In other words, the allocation of parking spaces is optimized according to the desire for them. At the same time, users of MonkeyParking must also pay any fees at their meters, so city authorities still get their money.

> Q: Is MonkeyParking a good design?

The service seems to open the way for forms of abuse such as drivers squatting in desirable spaces solely for the purpose of making money by leaving them. The developers argue that they can detect and punish such forms of abuse.

The argument that auctioning parking spaces reflects people's desire to have them assumes that bidders are well able to represent their desires with money. However, some people who would really like to have a particular parking spot may not have enough money to represent their desires adequately. A driver might want to bid $20 on a spot, for example, but lack the wealth to do so. A wealthier driver may outbid them, even though that person actually wants the space less. The design of MonkeyParking mitigates this problem to some extent by capping bidding at $20 per spot. Still, people with low incomes sometimes will see their desires under-represented in the final outcome of an auction.

Another issue with the design is that it appears to treat public parking spaces as the property of the service, one that users can buy and sell amongst themselves. In fact, the spaces at issue are owned by the cities where they are located. The developers counter that they are auctioning not the spaces but the information that the spaces will soon be vacant. However, this description seems incomplete. After all, people occupying spots that have just been auctioned are expected to squat in

---

[9]Cf. Xie (2014).

them until the auction winners arrive, thus excluding other drivers. Such exclusion seems indistinguishable from the exercise of property rights.[10]

The case illustrates the importance of social contracts to some design problems. As remarked by a columnist for The Economist magazine: "Are parking spaces the sort of thing we allocate through auction or are they the kind of thing that Sergey Brin and I have an equal chance at getting, even though he has a billion dollars and I don't?"[11]

Case Study: Airline Seats

Anyone who has flown in economy class on an airliner has probably experienced some discomfort with the seating, perhaps due to the small size of its design (Fig. 6). To address this issue, a lobby group called Flyers Rights has started a petition to have the US government create regulations to enlarge the minimum size of seats on US airlines.[12] Currently, airline seats in coach class vary from 17 to 19 inches in width (between the armrests) and from 31 to 34 inches in "pitch", the distance from one seat to the next one in front of it.

Seat measurements were originally based on US government survey of American body sizes in the 1960s. Seats were designed to accommodate people within the 95th percentile of hip width, meaning that only 1 out of 20 people would have hips that are too wide for their seats. Kathleen Robinette, who has studied body measurements over three decades for the US Air Force notes that the widest part of people's bodies are their shoulders, not their hips. Also, people have become larger, on average, than they were 50 years ago.[13]

So, with people increasing in size and seats tending to get smaller and closer together, they have become uncomfortable for more people. Furthermore, Robinette points out, seats that restrict movement can be unhealthy, since people who cannot move about are more at risk of conditions such as deep vein thrombosis.

However, airlines seats have become tighter due to economic pressures and not because airlines simply want to squash passengers. Commercial flights have become a bulk commodity, sold by online comparison engines on their relative cost. Sean Griffin of Boeing argues that economy class flyers have the following priorities (in order)[14]:

1. Flight availability at the time they want to fly
2. Cheap airfare

---

[10]The City of San Francisco, where the service was launched in the United States, deemed it illegal and ordered it to cease operations (Maddaus 2014).

[11]The Economist (2014).

[12]The Economist (2016).

[13]Patterson (2012).

[14]Hewitt (nd).

3. Marketing perks such as frequent flyer programs
4. Customer service issues
5. Comfort

To achieve the lower prices that passengers demand, airlines have to fit as many passengers on each flight as they can. As a result, seats have shrunk and been placed increasingly close together.

In this situation, the social contract is that commercial airplane coach seating is allocated essentially by auction. However, the result is seating that is uncomfortable and potentially unhealthy for many. In effect, the petition to mandate a larger, minimum size for coach seating would assign some priority to the needs of people who do not fit in it over smaller people for whom even current seating is not problematic.

Q: Do passengers have a right to larger seats?

One argument in favor of a minimum seat size is that it would help to save people from themselves. When people buy airline tickets, especially from an online service, considerations of cost and convenience are most prominent. They are displayed front-and-center on online services.

Considerations of comfort lie many weeks or months in the future, so there is a psychological tendency to discount them unduly. That is, people think of comfort not when buying tickets but only much later when they are actually stuck in uncomfortable seats. At that point, some people may wish that they had spent a little more money for a happier experience. Mandating a larger, minimum seat size would have the effect of helping people to fulfill that retrospective wish.

Some passengers already have a right to larger seats. For example, many airlines give a second seat to obese passengers, who simply do not fit in a single seat. Air Canada, for example, considers obesity to be a medical condition and thus gives a second seat to obese passengers who present a doctor's note. To demand a second fare from obese passengers would be a form of discrimination against people with disabilities, a violation of their rights to equality of access that would likely be punished as the result of a lawsuit if not respected.[15]

However, the same consideration is not granted to very tall passengers, who also do not fit in normal, coach seats. Airlines argue that tallness is not a medical condition, and so does not merit special consideration.[16]

---

[15]The Economist (2012).

[16]Steele and Hermiston (2015).

Fig. 6 An airplane cabin interior. Photo by Victor Toh. Detail of URL: https://pixabay.com/en/airplane-cabin-passenger-aircraft-734363/

Defaults

One response to the challenges of designing things in a way that respects people's rights is to make designs configurable by users. In that way, it is users and not designers who determine how a design behaves. To the extent that users determine what a design does, assessment of that design becomes an assessment of what individual users do with it.

This approach is quite reasonable, but leaves out a significant issue, namely that of *defaults*. A default characterizes how a system behaves in the absence of any modifications. Defaults are normally determined by designers and seldom questioned by users. As a result, most of the impact that designs have on users is still determined by the designs themselves.

Defaults can have substantial effects on how a system behaves. For example, consider the organ donation rates in these pairs of similar countries (Fig. 7)[17]:

In each pair, the top entry is a country where the default is that citizens are not enrolled in the organ donation program by default. In order to enroll, citizens need to explicitly join, or *opt in*, to the program. In each pair, the bottom entry is a similar country where the default is the reverse, that is, citizens are enrolled in the organ donation program by default. In order to leave the program, citizens need to explicitly un-enroll, or *opt out*.

Given the geographic and cultural similarities in these pairs of nations, it is clear that the default setting of the organ donation program is a crucial determiner of its performance.

---

[17]Johnson and Goldstein (2003).

Opt-in	Denmark: 4.25%	Netherlands: 27.5%	Germany: 12%
Opt-out	Sweden: 85.9%	Belgium: 98%	Austria: 99.98%

Fig. 7 Organ donation frequency in similar countries that differ in default enrollment, either opt-in or opt-out

Case study: Default Menus

An interesting example of the importance of defaults comes from a recent study of defaults in a children's menu at a restaurant chain called Silver Diner. Chef Ype Von Hengst, a co-founder of Silver Diner, thought that this design made the menu unhealthy since fries and soft drinks are not the best sort of food for children (or anyone else). In 2012, he decided to redesign his menu with healthier default side orders.[18] French fries and soft drinks were removed from the children's menu, although they could still be ordered by request. Healthy side items like salads and strawberries were made the default options in their place.

Researchers at Tufts University took an interest in this experiment and followed the results, now published in the journal *Obesity*:

> Before the changes, only about 3 percent of meals ordered off the children's menu qualified as healthy—meaning they met the nutritional standards set by the National Restaurant Association's Kids Live Well program. After the menu revamp, 46 percent of meals ordered met that standard.

> And while 57 percent of customers ordered French fries for their kids off the old menu, only 22 percent still requested fries after they disappeared from the menu. All told, about 40 percent of customers stuck with the default side dishes — regardless of whether the sides were fatty or healthy.

Patrons seemed not to mind the average 19-cent average increase in the price of the meals.

In eating, as in many other activities, people frequently stick with defaults. Since eating behavior is important to health, design of the default menu choice is a significant consideration for its designer.

Assessment of good design is not limited to designs merely as a means to an end. The concept of good design extends to the ends that a design serves. Moral concepts help in this form of assessment.

One important moral concept in this context is that of rights. A right is an entitlement to a certain kind of respectful treatment from others.

A social contract is a set of rights that regulate how people get along and share things. By entering into a social contract, people are better able to cooperate and thus to thrive. Since many designs direct or influence how people interact with and treat each other, the social contract concept can be applied in the moral assessment of those designs.

---

[18]Singh (2015).

In the following chapters, we will explore further the idea of a social contract and how it can be used in the moral assessment of designs.

References

Alphabet. (2015, October 2). *Alphabet code of conduct.* Retrieved January 26, 2016, from Google: https://investor.google.com/corporate/code-of-conduct.html

Churchill, W. (2008). *Churchill by himself: The definitive collection of quotations* (R. M. Langworth, Ed.). New York: Public Affairs.

Hewitt, E. (nd). *The shrinking airline seat.* Retrieved May 25, 2016, from Independent traveller: http://www.independenttraveler.com/travel-tips/travelers-ed/the-shrinking-airline-seat

Horn, P. (1995). *Children's work and welfare, 1780–1890.* Cambridge: Cambridge University Press.

Johnson, E. J., & Goldstein, D. (2003). Do defaults save lives? *Science, 302*(5649), 1338–1339.

Livingston, J. (2007). *Founders at work: Stories of startups' early days.* Berlin: Springer.

Locke, J. (1689/1988). *Two treatises of government* (P. Laslett, Ed.). Cambridge: Cambridge University Press.

Maddaus, G. (2014, September 18). *Kicked out of San Francisco, MonkeyParking app plans a fresh start in Santa Monica.* Retrieved May 25, 2016, from LA Weekly: http://www.laweekly.com/news/kicked-out-of-san-francisco-monkeyparking-app-plans-a-fresh-start-in-santa-monica-5080436

Patterson, T. (2012, June 1). *Airline squeeze: It's not you, 'it's the seat'.* Retrieved May 25, 2016, from CNN: http://www.cnn.com/2012/05/30/travel/airline-seats/

Selinger, E. (2014, February 26). *Today's apps are turning us into sociopaths.* Retrieved February 27, 2014, from Wired: http://www.wired.com/2014/02/outsourcing-humanity-apps/

Singh, M. (2015, April 29). *When you make kids' meals healthier by default, they still eat 'em up.* Retrieved May 6, 2015, from National Public Radio: http://www.npr.org/sections/thesalt/2015/04/29/403086469/when-you-make-kids-meals-healthier-by-default-they-still-eat-em-up

Steele, L., & Hermiston, S. (2015, February 9). *Too tall to fly? Airlines accused of height discrimination.* Retrieved May 25, 2016, from CTV Vancouver: http://bc.ctvnews.ca/too-tall-to-fly-airlines-accused-of-height-discrimination-1.2219269

The Economist. (2012, November 12). *How should airlines treat larger passengers?* Retrieved May 25, 2016, from The Economist: http://www.economist.com/blogs/gulliver/2012/11/obese-flyers

The Economist. (2014, July 12). *A bunch of new apps test the limits of the sharing economy.* Retrieved July 19, 2014, from The Economist: http://www.economist.com/news/business/21606874-bunch-new-apps-test-limits-sharing-economy-antisocial-networks

The Economist. (2016, February 16). *Legislation to make flying in America more comfortable has failed.* Retrieved May 25, 2016, from The Economist: http://www.economist.com/blogs/gulliver/2016/02/stretch-too-far

Xie, J. (2014, May 23). *A new parking app that's virtually guaranteed to stir up controversy.* Retrieved May 24, 2014, from The Atlantic CityLab: http://www.citylab.com/commute/2014/05/a-new-parking-app-thats-virtually-guaranteed-to-stir-up-controversy/371268/

Social Agendas

Abstract A social contract defines a body of rights that people respect in order to share resources and collaborate in a mutually beneficial way. Sometimes, social contracts are embodied explicitly in laws. However, social contracts are often implicit and may originate from many sources. One such source is designs themselves, whose designers have particular views about the social arrangements that people should operate under. Recall that Dieter Rams thought that designers should aim to bring about a humane world. Such views are often embodied in design movements that suggest an ideal world that designers should attempt to foster in their work. Modernism, as noted earlier, suggests that there is a universal, industrial lifestyle that good design should tend towards. Ideals like this may be called social agendas. A social agenda is not a set of cultural expectations that designs may satisfy but an ideal that designs promote for people to follow. Several examples of social agendas in design are discussed, but the list is open-ended.

Introduction

In the previous lecture, we began a study of the moral sense of good design. A moral evaluation of designs begins with consideration of the ends of those designs, that is, whatever outcomes they are supposed to achieve. In particular, we examined how designs may be assessed by evaluating their goals in the light of people's moral rights.

In the view of John Locke, rights typically come in packages that can be called social contracts. Social contracts embody sets of rights that prescribe how people ought to treat one another with the ultimate aim of promoting human thriving. When it comes to design, social contracts may specify the ends and the rights that apply to a particular technology. For example, the end or goal of a crosswalk is to allow pedestrians and drivers to share certain stretches of roadway. This social contract also defines a special right, a *right of way*, and how it is applied to enforce turn-taking in usage of the crosswalk. We can evaluate crosswalk designs, in part, by determining how well they facilitate this social contract.

© Springer International Publishing AG 2017   105
C. Shelley, *Design and Society: Social Issues in Technological Design*,
Studies in Applied Philosophy, Epistemology and Rational Ethics 36,
DOI 10.1007/978-3-319-52515-0_7

However, not all social contracts exist as sets of laws. Some are established simply by common understandings. Others are suggested by designers themselves. That is to say, designers sometimes act in the role of legislators. Their designs may promote one way of living over another way. Recall Dieter Rams's suggestion that his designs are intended to promote a lifestyle that is humane and without aggression or upset. His hope was to help users of his designs to live in this way.

In the role of legislators, designers may follow design movements that promote a particular social ideal. This ideal is a kind of picture of what society should be like, how people should treat each other, and so forth. This ideal can be called a *social agenda*. As such, we can evaluate designs by considering the social agenda that they promote. That is, whether or not a given design is a good one relates to whether or not the social agenda behind it is appropriate.

To carry out this form of evaluation, we need to characterize in more detail what a social agenda is and discuss some pertinent examples.

Case Study: The Juicy Salif

To clarify this notion, we may examine a design that exemplifies an important 20th century design movement, namely *postmodernism*. One design that illustrates the ideals of this movement and its social agenda is the Juicy Salif (Fig. 1) by French designer Philippe Starck (Fig. 2), produced in 1990. The Juicy Salif is a biomorphic/phallic/alien artifact with some modest functionality as a lemon-squeezer. Mostly, it is useful as a conversation piece or as testimony to the sophisticated tastes of the owner.

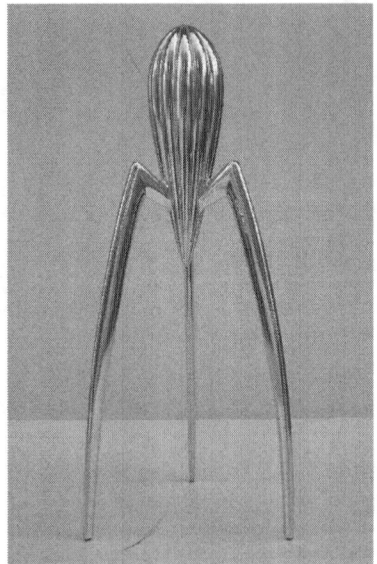

Fig. 1 The Juicy Salif, a lemon squeezer designed by Philippe Starck. Photo by Phrontis/Wikimedia commons. URL: https://commons.wikimedia.org/wiki/Category:Juicy_Salif#/media/File:Zitronenpresse_JuicySalif.jpg

In order to obtain lemon juice, users squash a half-lemon over the top, juice flows down the channels along the sides of the object and drips into a glass placed underneath. Guy Julier describes it this way[1]:

> As a utilitarian kitchen implement, it only half-works. It delivers lemon juice in an enticing, amusing fashion straight to a glass, but you also get the pips and some pith with it which may then need straining out. For the novice, juice splatters about: it is only after one learns the correct body posture, the optimum conjunction of limbs and the requisite force needed, does this problem recede. As if to underline this connoisseurship of function, the instructions supplied with the object include copious advice and detailed drawings on its use and cleaning. They also tell us that upon its first use, a chemical reaction takes place between the lemon juice and the aluminium, rendering the first squeezing redundant. Equally, the metal discolours, losing its shine.

As Julier notes, these difficulties of using the Salif for juicing were part of Starck's intention for the design. Here is how Starck himself describes how he sees the Salif being used:[2]

> This is not a very good lemon squeezer: but that's not its only function. I had this idea that when a couple gets married it's the sort of thing they would get as a wedding present. So the new husband's parents come round, he and his father sit in the living room with a beer, watching television, and the new bride and mother-in-law sit in the kitchen to get to know each other better. 'Look what we got as a present', the daughter-in-law will say.

In other words, it is more of a conversation piece, or a way for owners to show off their exotic taste in kitchen gadgets.

---

[1]Julier (2008), p. 75.
[2]Quoted in Julier (2008), p. 76.

Q: Is the Juicy Salif a good design?

As a lemon squeezer, the Juicy Salif cannot compare in function to the Braun Citromatic. Unlike the Citromatic, the Juicy Salif is meant to draw attention to itself and to stand out from the background. Starck seems more occupied with the social capital that owners of the Juicy Salif will gain than with any lemon juice they will get. No doubt Dieter Rams would regard it as a gimmick.

Postmodernism

The Juicy Salif illustrates the kind of design characteristic of postmodernism. Historically, this movement began as a reaction against modernism and against its minimalist and disciplinary rationality. There is no strict definition of postmodernist design. However, postmodernists typically promote:

1. Relaxation of rationality; *fun*
2. Engagement, spectacle, particular significance, branding; *whimsy*
3. Eclectic forms, irregularity, incongruity, unclear functionality.

The first item identifies the kind of social conditions that postmodernists regard as ideal. Where modernism is about calmness, order and discipline, postmodernism promotes exuberance, self-expression, and unrestraint. The second item identifies the design norms through which the social ideals may be realized. Postmodernists value pizzaz, evocation of personal and cultural associations in their designs, and also associations with the designers themselves. The third item identifies how these norms are carried out. Postmodernist designs often exhibit components from different historical periods, are irregularly arranged, and resist easy or casual understanding of their function.

Q: In what ways is the Juicy Salif postmodernist? What other examples can you think of?

Agendas   *Social agenda: ideals of how society should work*

There are many different kinds of social agendas in design. Some well-known examples include modernism, feminism, and consumerism. Modernism involves the view that people's goods and spaces should be organized along rational principles that apply equally to everybody everywhere. This can be achieved in designs through minimizing and mass production. Feminism involves the view that society should be organized to eliminate discrimination against women. Feminism can be a political agenda where it advocates for political changes in favour of women. (It can also be a professional agenda, where it advocates for more representation of women in engineering, for example.) Consumerism involves the view of people as consumers of services and resources, as opposed to producers of them, for example. This agenda often involves promotion of consumption and the protection of consumers, e.g., consumer safety.

> Q: Which design agenda is better: Modernism or postmodernism?

For present purposes, let us define a social agenda in design as an ideal for how people ought to live and interact, consisting of the following components:

1. Vision: An image of ideal social conditions to be established,
2. Values: Norms for design that promote the vision, and
3. Methods: Methods for how the values should be implemented.

Sometimes, design agendas are equated solely with the third component, that is, with how designs are presented by promoters of a given agenda. However, social agendas in design are more than simply ways of styling products.

Genderism

The architectural historian Adrian Forty provides a detailed discussion of social agendas among British designers in modern history (Fig. 3).[3] For example, Forty notes that the distinction between *masculine* and *feminine* is a very important agenda in British (and other) societies.[4] In other words, societies take the physical differences between males and females and incorporate this difference into a broader distinction about what is masculine and feminine.

These distinctions include the design of various goods such as watches. Forty points out that men's and women's watches in Edwardian Britain exhibited some systematic differences. For example, at one price point, men's watches tended to be larger and more robust. Ladies' watches were more ornate and featured more delicate looking cases and hands. In the 1907 Army and Navy Stores catalogue, men's watches were all calibrated with Roman numerals whereas women's watches were all calibrated with Arabic numerals. Evidently, the designers felt that the stiff

---

[3]Forty (1986).
[4]Forty (1986), pp. 65–66.

and erect appearance of Roman numerals made them more suited to men while the curved and thin appearance of Arabic numerals made them more suited to women. In other words, gender in watches is reflected at least in size, ornament, and font. As suggested by the design of these watches, the agenda of genderism involves:

1. Vision: A broad distinction between male and female persons.
2. Values: A distinction between masculine and feminine things.
3. Methods: Size, robustness, ornamentation, etc.

Q: How do other designs differ by gender today? When is genderism in design appropriate?

Genderism may be appropriate where men and women tend to have different requirements for certain designs. For example, power tools with smaller handles may be more appropriate for women, who tend to have smaller hands than do men.

However, genderism may be inappropriate where it stifles appropriate access to resources. For example, having men's and women's restrooms may discriminate against transgendered persons, who might be uncomfortable in one restroom and unwelcome in the other.

Note that genderism is not the same as *sexism*. Sexism presupposes genderism but its vision is (typically) that men are better than women. Thus, it promotes the value that masculinity is *superior* to femininity. Genderism assumes that the sexes are different but does not assume any inequality between them.

Fig. 3 Adrian Forty, Professor of Architectural History at the University College London, whose book, *Objects of desire*, explains relationships among consumers, designers, and their goods. Photo courtesy of Adrian Forty

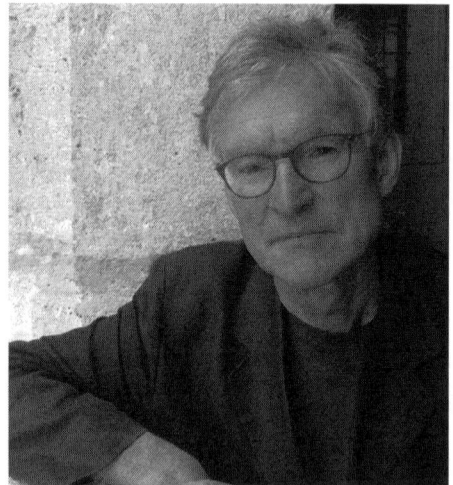

Mythologizing

Of course, genderism is not a social agenda invented by designers. Instead, it was a part of prevailing British culture that designers sought to comply with. As such, it may seem to be only a kind of contextualism, that is, the practice of making designs that fit with their cultural context.

However, Forty argues that there is more to genderism than simple compliance with cultural norms. Besides adhering to cultural norms, genderized designs also tend to *reinforce* them. This reinforcement happens through a process that he refers to as *mythologizing*. A design mythologizes when it makes a cultural practice seem more like a law of nature.

As Dieter Rams pointed out, we increasingly live in a world of our own design. When social ideals such as gender become embedded in the design of that world, it can begin to seem as though that ideal, the distinction between masculine and feminine in this case, is just a fact about the world that people have discovered instead of something that people themselves have invented. This tendency of social agendas increasingly to appear "built in" to nature is what is referred to as mythologizing.

One implication of mythologizing is that there is often a feedback between cultural ideals and features of design. Consider how gender is represented in clothing. Clothes are often designed to be masculine or feminine. Men and women generally wear clothing that visibly conforms to this ideal. In turn, the fact that the distinction is observed so regularly tends to reinforce it. That is, the more that people conform to the ideal, the more compelling that ideal becomes.

This feedback relation may be illustrated as in Fig. 4.

So, a social agenda in design is more than simply compliance of designers with prevailing cultural ideals. It also involves the reinforcement of ideals by making them more real and, thus, seemingly natural and inevitable.

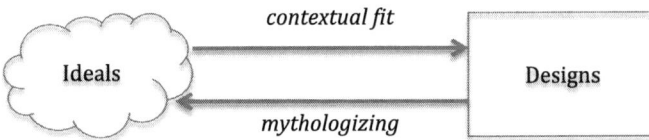

Fig. 4 Mythologizing and contextual fit as reciprocal constraints between designs and cultural ideals

Fig. 5 Universal Electric
Mixabeater, ca. 1920. Photo
by Tom Rent. URL: http://i.
ebayimg.com/images/g/
acIAAOSwzrxUvoRg/s-
l1600.jpg

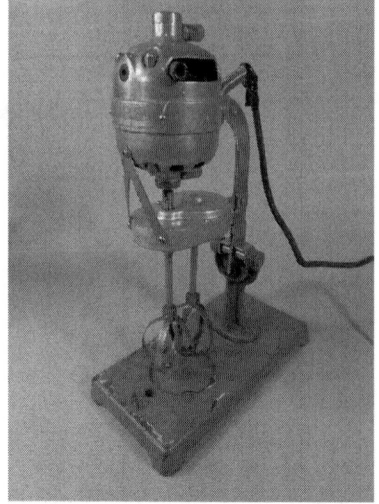

Design for Housework

An interesting example of mythologizing discussed by Forty concerns housework.[5]
In 19th century Britain, housework was often done by servants. Wealthy households
had a staff of servants to do the cooking, cleaning, and washing up. Middle-class
households might employ one or two servants to do the dirty work.

In the early 20th century, the number of middle-class households had expanded
but the population of servants had not. This situation meant that there was a sizeable
increase in the amount of housework to be done but no increase in the labor pool to
accomplish it. Since the men of British households were expected to work outside
of home, the task of performing the extra work fell to women. Because this work
was associated with lower-class servants, it was considered socially demeaning for
these women of the middle class.

A resolution of sorts to this issue came with some industrial design. Small
appliances that were used to perform housework, such as washers, dryers, vacuum
cleaners, and mixers, were redesigned to appear less like industrial machines and
more like furniture of a sort.

For example, the first models of electric food mixers designed for household use
appeared around 1920 and looked much like machines that might be found in a
factory.[6] Early household appliances were often simply smaller and less powerful
versions of equipment already found in factories and other commercial settings. See
Fig. 5.

---

[5]Forty (1986), pp. 207ff.
[6]Forty (1986), pp. 216–217.

Consumerism

Big refrigerators are associated with the social agenda of *consumerism*. In this sense, consumerism involves the simple promotion of consumption. (Maybe it should be called *consumptionism*.) This sense may be described as follows:[12]

1. Vision: Consumption as a crucial economic activity.
2. Values: Cheapness, disposability and disengagement.
3. Methods: Ephemeral materials, uninteresting appearance, sealed workings.

Some things are designed to suggest their disposability to users. Paper cups, pop cans, and plastic bottles are typically small, flimsy items that are quickly used up and not readily cleaned and re-used. Thus, the most obvious thing to do with them is to throw them away, and not always in a trash can. After all, when another one is needed, it will be cheap and readily available.

Even solid and sizeable items can be designed for consumerism. For example, refrigerators are not designed to be easy or cheap to maintain or repair, prompting owners to throw them out rather than fix them when they break down.

Where designs themselves do not prompt consumerist views of goods, advertisers may design ads to do so. A classic example is the IKEA lamp ad of 2002, directed by Spike Jonze. In the ad, the idea that a desk lamp is something that should be valued and used up is held up to ridicule. Instead, the act of throwing away a working desk lamp is praised as an exciting thing to do. This attitude then encourages people to be lax about replacement of goods that remain workable simply for reasons of novelty or fashion.[13]

The idea of the ad is to encourage people to view desk lamps in the same way they view paper cups. The hope is that they will then be more likely to buy an Ikea desk lamp, even if they already own a lamp that works.

Case Study: The Bic Pen

A good example of an item designed for the consumerist agenda is the Bic pen (Fig. 7).[14] The pen was invented by Marcel Bich in France after WW II. The story has it that he was inspired when using a wheelbarrow to imagine how a generic ball-point pen would work. Essentially, Bich figured out how ink could be loaded to flow from a tube through the ball point without blotting or allowing air back into the tube. He also designed the process needed to turn out the pen in bulk.

---

[12]Cf. Crawford (2015).

[13]Hales (2002).

[14]Lidwell and Mancasa (2009), pp. 54–55.

Fig. 7 A Bic Cristal pen, one
of hundreds of billions. Photo
by Trounce/Wikimedia
commons. URL: https://
commons.wikimedia.org/
wiki/File:03-BICcristal2008-
03-26.jpg

In 1950, Bich founded a company (later *Société Bic*) that manufactured the pen and also shortened the name to avoid any unfortunate confusion with the English word "bitch".

The pen has many good functional attributes. The ball is very sturdy and resists wear, thus allowing it to draw a uniform and steady line. Ink flows appropriately from the reservoir through the tip and not the other way around. The hexagonal shape facilitates grip and prevents the pen from rolling away on desktops. The colour-coded cap and butt allows users to easily see what colour ink the pen contains.

By the same token, the Bic is a throw-away design. It is slight and insubstantial and does not lend itself easily to refills or reservoir replacement. It is often purchased in bulk, further signaling that it is not meant to be maintained or kept around. And, it is cheap.

According to Guinness World Records, Bic is the world's best-selling pen, having sold its 100 billionth pen in 2006. The pens sell at a rate of about 57 per second worldwide.[15]

Q: What other examples of consumerism in this sense can you think of? Is consumerism a good design agenda?

Environmentalism

By contrast with consumerism, environmentalism is a social agenda that aims at restraining consumption or, at least, some of its consequences. This agenda could be described as follows:

1. Vision: Integrity of the natural environment.
2. Values: Pollution reduction, environmental awareness.
3. Methods: Reusability, biodegradability, etc.

---

[15]Guinness World Records (2006).

One form of pollution is litter. Littering occurs when people improperly dispose of their goods, e.g., candy wrappers, coffee cups, cigarette butts, etc (Fig. 8). Litter is a form of pollution that can threaten the integrity of ecosystems. Plastic shopping bags, for example, break down in the environment into bits of plastic that are eaten by animals, interfering with their digestion and sometimes choking them.

One approach to environmental preservation is to regard litter as an issue of attitude. That is, litter is caused by "litter-bugs", people who are lazy or inconsiderate in their disposal of used goods. One solution to the litter problem on this agenda was an education campaign to get people to change their behaviour. Perhaps the best-known instance centered on the "crying indian ad" of the 1971 Keep America Beautiful campaign.[16]

This approach has been criticized on two grounds. First, it seems based on the image of littering as only a behavioural issue, and thus addressable by an educational campaign. However, littering is also driven by the emphasis on disposable goods in society, our "disposable culture".

Second, the campaign was sponsored by companies such as Philip-Morris, Anheuser-Busch, PepsiCo, and Coca Cola, which sell large quantities of disposable goods that end up as litter. Thus, the campaign may be seen as a method for avoiding responsibility as much as dealing with the litter problem.

Another approach to litter regards it as a motivational issue. In other words, people are not normally motivated to dispose of garbage properly. Gamification attempts to take a tedious or low-reward task and turn it into an amusement.[17] In one case, the so-called "fun theory" is applied to change the nature of litter collection from a tedious duty to an engaging form of play.

A simple example would be the "World's deepest bin", a garbage can that produces a sound effect making it seem that garbage tossed into the can falls into a deep pit before finally hitting bottom.[18] Informal tests of the design found people would make more of an effort to dispose of trash in the bin in order to enjoy the sound effect that the bin produced as a result.

The effect of this design is that it prompts people to pick up and dispose of garbage simply in order to hear the sound effect. The result is that the ground near the garbage can is much cleaner than the ground around ordinary cans.

Q: How does this design compare to the ad campaign?
Q: Which agenda is better: Consumerism or environmentalism?

The ready availability and omnipresence of disposable goods suggests to people that simply tossing things away is an acceptable practice. Designers of goods and services could design them in a way that discourages this view.

---

[16]Dunaway (2015), pp. 79–95.

[17]Deterding et al. (2011).

[18]Volkswagen (2009).

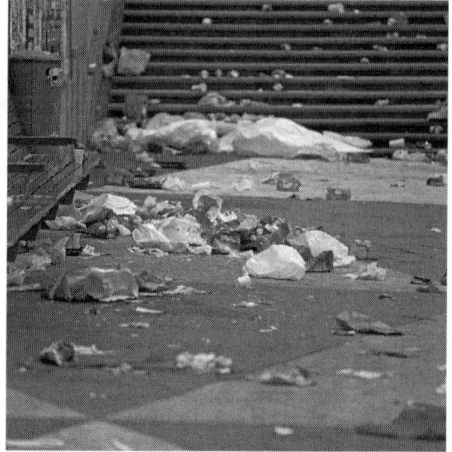

Technology Solutionism

Technology solutionism is a social agenda in design identified by technology critic
Evgeny Morozov (Fig. 9).[19] In his criticism, Morozov characterizes the social
agenda of much of the information technology sector as an effort to undermine (or
"disrupt") important social institutions and replace them with privatized services.
Technology solutionism could be characterized as follows:

1. Vision: Undermining government through technology.
2. Values: Privatization of social institutions.
3. Methods: Personalization, "app-ification".

Consider the following example from Morozov, in which he refers to technology
solutionism as the "open agenda":[20]

> "The open agenda is, in many ways, the opposite of equality and justice. They think
> anything that helps you to bypass institutions is, by default, empowering or liberating. You
> might not be able to pay for health care or your insurance, but if you have an app on your
> phone that alerts you to the fact that you need to exercise more, or you aren't eating
> healthily enough, they think they are solving the problem."

The health app example illustrates Morozov's concerns. A smartphone health
app that prompts users to exercise may help them become healthier (or not).
Morozov worries that it may also weaken the resolve to maintain a robust, public
health care system. A social health care system exhibits "equality and justice" in the
sense that it treats people according to need rather than according to ability to pay.

---

[19]Morozov (2013).
[20]Packer (2013).

Fig. 9 Evgeny Morozov,
author of *To save everything,
click here*, and critic of
technological solutionism.
Photo by January/Wikimedia
commons. Detail of URL:
https://commons.wikimedia.
org/wiki/Category:Evgeny_
Morozov#/media/File:
Evgeny_Morozov,_Author,_
To_Save_Everything,_Click_
Here_(8568053409).jpg

However, in the presence of health apps that nag individuals to stay healthy, many people may lose interest in funding a collective health scheme.

There are two reasons to resist such a disruption, according to Morozov. First, a more individualized system of health care will mean that when people in poverty get sick, they will be unable to access services that they require and that they could access in a social scheme. Since access to health care is a basic right, in his view, then this outcome is unjust.

Second, an individualized and privatized system of health care might well cost more, overall, than a social system while providing a similar, overall outcome. The difference will be that poorer people will be sicker and wealthier people will be healthier.

In Morozov's view, some tech designers employ technology solutionism in a cynical way, that is, they knowingly disguise their aim to undermine social institutions in the form of glitzy apps that promise personal empowerment. Other tech designers may be merely misguided, unaware that their designs may have anti-social effects.

Case Study: Waze

Let us explore how this idea might be applied to other, actual designs. Consider the example of an app-based service named Waze. People with the app can use it to receive turn-by-turn directions for car navigation. In addition, Waze includes up-to-date traffic conditions in its directions and attempts to route users in the quickest way possible to their destinations.

There is no doubt that people like getting places quickly and dislike being stuck in traffic. Moreover, if many people use Waze, it might help to relieve congestion in a city on a large scale. Nevertheless, Morozov might still consider the service to be an example of technology solutionism.

Q: How might Waze be an instance of technology solutionism? Is it a good design?
Q: Can you think of other examples of technology solutionism?

The vision embodied by Waze seems to be easing of traffic congestion. Its main value is conceptualizing congestion as a matter of routing, and its main method is the personalization of driving directions to individual drivers. Morozov would likely argue that traffic congestion is due to many social issues, such as road quality and design, traffic control, and public transit, all of which are hidden by the service.

Interestingly, Waze has begun partnerships with several governments, such as the city of Rio de Janeiro and the State of Florida.[21] In this deal, governments provide Waze with information about potential traffic disruptions, e.g., construction, in exchange for government access to the Waze real-time database for "planning purposes." For example, the city of Rio de Janeiro is using Waze data to plan garbage truck routes.

Not all government agencies are happy with this development. For example, the National Sheriff's Association wants Google to remove the function that allows users to report the locations of police cars. Primarily used to report the location of speed traps or construction sites, the Association worries that it could be used to plan attacks on police. Some police officers use the function to misreport their locations, something that the app is programmed to catch and dismiss.

Critics are also concerned that the connection will mean governments will structure traffic control in favour of Waze users, to the detriment of other drivers and the users of other routing services.

Taylorism

Following the Industrial Revolution, increases in efficiency of production became an important concern for manufacturers. At first, attention focused mainly on the invention of new equipment that could help to turn out goods faster or with less work.

However, the design of manufacturing work itself became a focus of attention. Around the turn of the 20th century, Frederick Taylor began systematic study of industrial labor.[22] Taylor considered industrial workers to be naturally lazy, working as little as they could get away with. He noted that many activities performed by workers were not obviously productive, e.g., smoking or chatting with co-workers. He and his followers used stopwatches and slow-motion photography to study how manufacturing labor was performed. They also consulted with

---

[21]Ungerleider (2015).
[22]Saval (2014), pp. 45ff.

manufacturers to design more efficient methods of labor. The design of manufacturing facilities, and the movements of workers themselves, were changed and regimented to increase productivity.

The social agenda pursued by Taylor has become known as *Taylorism*. Taylorism may be described as follows:

1. Vision: Increased productivity of manufacturing.
2. Values: Efficiency, top-down organization.
3. Methods: Automation, deskilling, mass production.

Prior to Taylorism, it was often assumed that the way to get some product manufactured was to train workers with the skills and expertise to make it. Then, they could simply make the product using supplies kept on hand.

Taylor realized that it would be cheaper to use unskilled labor and transfer the knowledge and skills needed to make a product to managers and the manufacturing process itself. For example, instead of assembling cars by hand, it would be more efficient to use an assembly line. On an assembly line, each worker's job could be kept quite simple, e.g., tightening some nuts on an engine. In such a low-skill system, no worker knows how to make a car. Instead, that knowledge is embodied in the complex process of assembly itself.

The substitution of unskilled labor for skilled labor is known as deskilling.

Case Study: The Efficiency Desk

In the Victorian era, office work became the quintessential, middle-class occupation. Special desks were designed to help office workers with the crucial task of handling paperwork. This sort of desk is epitomized by the Wooton Secretary Desk. See Fig. 10. It contained dozens of drawers, slots, pigeonholes, and boxes, all to allow users to store different sorts of written materials close at hand. The sides of the desk could be swung in front of the desk so that the whole assembly could be locked up like a safe when not in use.

The design of the desk suggests that Victorian office workers were often trusted to manage their own work, which they kept right at their own desk under their own control. Each office worker was a fairly autonomous individual who, it was assumed, possessed the knowledge and skills needed to do their work.

Like manufacturing work, this situation was reconsidered by Taylorist efficiency consultants. W.H. Leffingwell, a protégé of Taylor, studied office work with the object of making the flow of paperwork more efficient. As with manufacturing work, the result was to remove much of the autonomy of office workers. Office spaces were organized less like domestic rooms and more like factory floors, with a grid of desks sitting in an open area, which could be supervised from a platform or balcony by a manager, as in the Larkin Building designed by Frank Lloyd Wright.

Paperwork itself was stored not in individual desks but in centralized filing cabinets to which any worker might have access. In this way, paperwork would

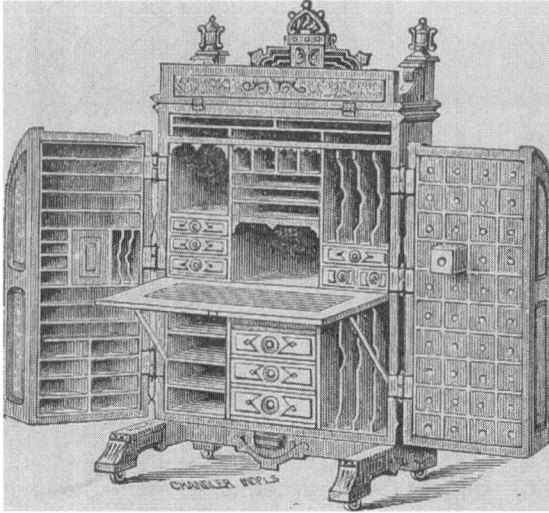

Fig. 10 A Wooton secretary desk. Picture courtesy of the United States Library of Congress. URL: https://flic.kr/p/ocSUkE

flow smoothly throughout the office system, under the control of managers, rather than getting pigeonholed in some individual's desk.

As a result, the office desk itself became highly simplified. It became a simple, rectangular surface with a modest drawer or two, adequate to hold only a few office supplies and perhaps a bagged lunch. Appropriately, the new kind of office desk became known as the *efficiency desk*.

See Fig. 11. As Galloway (1919, p. 89) puts it, "As there is no room for placing current work in the drawers, any tendency to defer until tomorrow what can be done today is nipped in the bud."

Fig. 11 The efficiency desk

> Q: In what ways does the efficiency desk reflect Taylorism in office work?

In general, Taylorism had the advantage of making manufacturing or workflow more efficient, thus increasing productivity. This increase, in turn, helped to lower prices for products, making producers more competitive. At the same time, Taylorism tended to make work itself more repetitive and tedious. It also concentrated power in the workplace in the hands of a few managers, sometimes leading to labor unrest. Today, the term Taylorism is still associated with regimented forms of work organized or controlled through centralized management.

References

Crawford, M. (2015). *The world beyond your head: On becoming an individual in an age of distraction.* New York: Farrar, Straus and Giroux.

Deterding, S., Dixon, D., Khaled, R., & Nacke, L. (2011). From game design elements to gamefulness: defining "gamification". *Proceedings of the 15th International Academic MindTrek Conference: Envisioning Future Media Environments* (pp. 9–15). New York: ACM.

Dunaway, F. (2015). *Seeing green: The use and abuse of American environmental images.* Chicago: University of Chicago Press.

Forty, A. (1986). *Objects of desire: Design and society since 1750.* London: Thames and Hudson.

Galloway, L. (1919). *Office management: Its principles and practice.* New York: The Ronald Press Company.

Guinness World Records. (2006, Sep 1). *Pen—best selling.* Retrieved Feb 1, 2016, from Guiness World Records: http://www.guinnessworldrecords.com/world-records/pen-best-selling/

Hales, L. (2002, Sep 28). Throw old stuff away! New ads deride attachment to worn-out things. *Washington Post.*

Hoffman, J. S. (1999). Appliances, energy, and the environment: The scale of the issue. In P. Bertoldi, A. Ricci, & B. H. Wajer (Eds.), *Energy efficiency in household appliances* (pp. 7–13). Berlin: Springer.

Julier, G. (2008). *The culture of design* (2nd ed.). London: SAGE.

Lidwell, W., & Mancasa, G. (2009). *Deconstructing product design.* Beverly, MA: Rockport.

Morozov, E. (2013). *To save everything, click here: The folly of technological solutionism.* New York: Public Affairs.

Packer, G. (2013, May 27). *Change the world: Silicon Valley transfers its slogans—and it's money —to the realm of politics.* Retrieved Oct 2, 2015, from The New Yorker: http://www. newyorker.com/magazine/2013/05/27/change-the-world

Rees, J. (2013, Oct 4). *The huge chill: Why are American refrigerators so big?* Retrieved Oct 1, 2015, from The Atlantic: http://www.theatlantic.com/technology/archive/2013/10/the-huge-chill-why-are-american-refrigerators-so-big/280275/

Rosenfeld, A. H. (1999). The art of energy efficiency: Protecting the environment with better technology. *Annual Review of Energy and the Environment, 24,* 33–82.

Saval, N. (2014). *Cubed: A secret history of the workplace.* New York: Doubleday.

Ungerleider, N. (2015, April 15). *Waze is driving into city hall.* Retrieved May 20, 2015, from Fast Company: http://www.fastcompany.com/3045080/waze-is-driving-into-city-hall

Vanek, J. (1975, Nov 1). Time spent in housework. *Scientific American*, 116–120.
Volkswagen. (2009, Sep 21). *The world's deepest bin: The fun theory*. Retrieved Feb 2, 2016, from TheFunTheory.com: http://www.thefuntheory.com/worlds-deepest-bin
Wansink, B. (2013). *Slim by design: Mindless eating solutions for everyday life*. New York: William Morrow.

Activism

Abstract As noted in the discussion of social agendas, good design sometimes implies designs that legislate, as it were, a social contract. Some social agendas are distinguished by their activism, that is, their aim to change the existing society in a significant way. Activism in design often displays a pursuit of social justice, that is, a society in which everyone receives their due and no one is wrongfully deprived of it. An activist design might seek to protect people in a vulnerable or disadvantaged social group, for example. There are many kinds of activism, e.g., guerilla or vigilante activism, humanitarianism, and social entrepreuneurism. Each kind involves a different view of what social groups are vulnerable or disadvantaged and how they might be best helped. Even though activism in design may be well intended, there may be circumstances in which it is inappropriate.

Introduction

In our discussion of social agendas of designers, we noted how designers sometimes act in a social role like that of legislators. That is, they design things in line with one or another social agenda. There are many such social agendas, including modernism, postmodernism, genderism, consumerism, environmentalism, etc.

Thinking about social agendas of designers can assist in assessment of their works. Each social agenda involves a kind of social contract, an ideal describing how people may thrive and get along with one another. To assess a design, then, we may consider how well the social agenda underlying it facilitates cooperation and thriving.

One feature that stands out about some design social agendas is their *activism*. Whereas social agendas of designers typically operate within the existing social order, some design agendas aim to change society in some significant way. Postmodernists and environmentalists, for example, might see themselves as effecting fundamental change in how people live.

Activism describes social agendas that are aimed at revising the prevailing social order. More particularly, activist agendas are aimed at changing the prevailing

© Springer International Publishing AG 2017

C. Shelley, *Design and Society: Social Issues in Technological Design*,
Studies in Applied Philosophy, Epistemology and Rational Ethics 36,
DOI 10.1007/978-3-319-52515-0_8

social order in order to achieve *social justice*. In broad terms, a just society is one in which each group in society receives what is due to it and is not deprived of what is due to it. Activist designers seek to apply their design skills to improve situations in which social justice is lacking.

Broadly, in terms of a social agenda, design activism may be described as follows:

1. Vision: Social justice increased.
2. Values: Accessibility, affordability, universality…
3. Methods: Collaborating with a marginalized clientele …

This description is vague because it is highly generalized. In this chapter, we will explore designer activism by looking at examples of how designers may seek to bring about social justice as they see it. Those ways include guerilla activism, humanitarianism, and social entrepreneurism.

Case Study: Knee Defender

We have already discussed the challenges of airplane seats in economy class becoming smaller and closer together. The group Flyers' Rights advocates for regulations establishing minimum seat sizes and spacing. Another approach would be to equip passengers to act individually in their own interests. This is the aim of the designers at Gadget Duck, who have invented the Knee Defender.[1] This device consists of two clips that passengers can attach to a tray table attached to the seat in front of them. Since the arms of these tray tables are integrated with the seat reclining mechanism, reclining can be restricted by restricting movement of the tray table. The designers of this device explicitly note on their web site their motivation of righting an injustice:

> It helps you defend the space you need when confronted by a faceless, determined seat recliner who doesn't care how long your legs are or about anything else that might be "back there".

> For those of us who have to squeeze ourselves into the limited airplane legroom space of a coach seat offered by many airlines, a seat in front of us that is poised to recline is a collision waiting to happen—with our knees serving as bumpers.

The device does not seem to contravene any FAA regulations, although flight attendants have been instructed to discourage their use due to their obvious potential to create conflict among passengers.

Q: Is Knee Defender a good design?

---

[1]Gadget Duck.

Guerilla Activism

The Knee Defender is an example of a form of activist design that may be called *guerilla activism* or, perhaps, vigilante activism. This term is used to capture the fact that such designers aim to equip people to, as it were, take the law into their own hands in order to address social injustices. As such, guerilla activism might be characterized as follows:

1. Vision: Empowerment of vulnerable people.
2. Values: Subverting established practice.
3. Methods: Alteration, sabotage …

As this characterization suggests, guerilla activism is frequently a matter of reacting against an established way of doing things, as represented by the way goods like aircraft seats are designed to function.

Q: What are other examples of guerilla activism?

Case Study: The MOM Incubator

Since 2011, Syria has been in the throes of a civil war. The war has created a grave humanitarian crisis within the country and a refugee crisis outside of it. Naturally, the situation presents many challenges. James Roberts was casting about for a design project for his senior year in the Product Design and Technology program at Loughborough University when he saw a TV documentary about some of the consequences of the war in Syria.[2]

> I was inspired to tackle this problem after watching a documentary on the high death rate among premature babies in refugee camps. It motivated me to use my design engineering skills to make a difference. Like many young inventors, there have been struggles along the way – I had to sell my car to fund my first prototype! The dream would be to meet a child that my incubator has saved—living proof that my design has made a difference.

Many deaths have been caused by lack of incubators due to their great expense. They can cost as much as $40,000. Mr. Roberts set out to design one that would be affordable to hospitals in the region of Syria.

The MOM incubator has many special features. The device can be collapsed for transportation and run from a battery that lasts 24 h, in case of power outages. The incubator is inflated manually and it is heated using ceramic heating elements.

---

[2]James Dyson Foundation (2014).

A screen shows the current temperature and the humidity, which can be varied depending on the baby's age. An alarm sounds if the temperature goes out of the desired range. For babies that suffer from jaundice there is a phototherapy unit.

Since the unit is inflatable, it is collapsible and compact. Also, air makes a good insulator, thus enhancing the function of the incubator without the use of expensive materials. The MOM incubator can also be made for under $400 per unit, two orders of magnitude less expensive than conventional designs. The MOM incubator won the 2014 James Dyson Award.

Q: In what ways does MOM serve social justice?

As very young children, Syrian infants are at a particular disadvantage through no fault of their own. This disadvantage is compounded by the poverty in which they are born, denying them access to medical care, which is designed to be unnecessarily expensive. Affordability of the MOM incubator exhibits the greater universalism of the design. That is, it is designed to give aid to as many people as possible and not only those who can afford expensive care or equipment.

Humanitarianism

Guerilla activism typically involves middle-class people in developed nations designing things for other people like them. Designs like the MOM incubator typically involve people designing for others who have much less money at their disposal. This sort of design agenda may be called *humanitarianism* because it is done for clients who usually cannot pay for the services of professional designers.

Humanitarian design may be characterized as follows:

1. Vision: Greater equality for people in poverty.
2. Values: Affordability, appropriateness.
3. Methods: Inexpensive materials, low energy requirements, …

From a professional perspective, it seems odd to want to work for people who cannot pay for the service. However, humanitarian designers often view people in poverty as being trapped in their unfortunate situation and still deserving of professional design work due to their needs.

For example, architect Samuel Mockbee ran the Rural Studio, a program where architecture students at Auburn University could design and build structures for poor African Americans in rural Arkansas. Mockbee noted the contrast between his students, who were white and affluent, and his clients, who were black and lived below the poverty line. Although these clients were unable to pay for the services of architects, Mockbee remarked of their situation that, "it's economic poverty, not moral poverty."[3]

Emily Pilloton

A good example of a humanitarian designer is Emily Pilloton (Fig. 1). Pilloton received her BA in Architecture from UC Berkeley in 2003 and a Master's degree in Product Design from the School of the Art Institute in Chicago in 2005. With these qualifications, Pilloton could have pursued a lucrative design career. However, she began to doubt the priorities represented by her profession[4]:

> At graduate school, people were starting to talk more about sustainability, but I felt it lacked a human factor. Can we really call $5000 bamboo coffee tables sustainable?

Bamboo is a sustainable material in the sense that it is grows quickly and is renewable. However, as Pilloton implies, sustainability usually means instigating a broad change in consumption patterns, which can hardly be brought about through designing and selling boutique furniture.

In 2007, she founded Project H (where "H" stands for "Humanity, Habitats, Health, and Happiness"), a non-profit group of designers who work on humanitarian projects. Projects undertaken by this group include the Learning Landscape (to which we will turn below), the Hippo Roller, and Studio H. Studio H is a design-build program for high-school students. To establish the program, she and designer Matthew Miller moved in 2010 to Bertie County, an impoverished region of North Carolina. There, they taught students to design and build structures ranging from chicken coops to the Windsor Farmers' Market. After financial support was withdrawn by the county in 2012, she moved the program to Berkeley, California.

Clearly, a significant difficulty in humanitarianism is acquiring enough funding to carry out projects whose beneficiaries do not have a great deal of money themselves.

In her talk to the PopTech 2009 conference, Pilloton lays out the central ideas of her approach to humanitarian design.[5]

Q: What central ideas guide Pilloton's approach to design?

To answer this question, consider the significance of the following four statements that Pilloton makes in her presentation:

---

[3]Cf. Dean (2002).

[4]Rawsthorn (2009).

[5]Cf. PopTech (2009).

Fig. 1 Emily Pilloton, founder of Project H Design, presenting at the 2009 Pop! Tech conference. Photo by Kris Krüg. Detail of URL: https://flic.kr/p/79vh66

1. "Design, I thought, was about problem solving and yet I wasn't getting to look at any of the big problems."
2. "We start at the beginning and not with the result."
3. "We work with and not for."
4. "We like to start locally ... but the ultimate goal is global scalability and adaptability".

Case Study: The Learning Landscape

Consider the *Learning Landscape* project designed by Emily Pilloton and other members of Project H.[6] The Learning Landscape is a grid of old tires, often 5×5 and half buried in sand, that are used as a platform for game playing. See Fig. 2. Simple games include "Match me", in which competing teams of children solve simple math problems by being the first to locate and sit on the tire on which the correct answer has been written in chalk. Project H has supplied a Web-based service where teachers can post and discuss other educational games that they have developed for the Learning Landscape.

Q: In what ways is the Learning Landscape humanitarian and not commercial?

One of the main design goals of the Learning Landscape is affordability. It emphasizes use of widely and cheaply available resources, namely, old tires and

[6]Pilloton (2009), p. 161.

Fig. 2 Drawing of the Learning Landscape, a grid of used tires, half buried in the ground, that can be used as a platform for fun, educational activities

manual labor. It also is clearly not geared to make money for Project H, since the project does not sell the Learning Landscape as a product. Instead, Pilloton describes the Learning Landscape as an example of "humanitarian design".

Social Entrepreneurism

A related form of activism in design is called *social entrepreneurism*, or social innovation. Proponents of social entrepreneurism aim to improve life for people in poverty and seek to do so through the application of design expertise. However, their approach is to adapt the model of entrepreneurial innovation from developed countries to the facts of life in developing ones. Not surprisingly, this approach raises some challenges.

As entrepreneurs, these activists rely on the marketplace to distribute their products. One reason for this approach is that they reject the usefulness of charity as a way of alleviating poverty. This is not because people who receive charity become lazy or dependent, but because they are not necessarily engaged with what they are given. Here is how Paul Polak of IDE explains the matter in a recent interview[7]:

> In Zimbabwe, for example, we did an experiment where we provided small irrigation drip systems to poor people through existing non-profit organizations, organizations that, by the rules of the donor, had to give them away. Only 25 percent of those drip systems were ever used. A lot of the people who accept things as a gift are not motivated to use them.

> Not expecting the poor to invest in their own wealth creation is a tragic mistake. They need to be willing to invest their own time and money. Great ideas are worthless if the farmer isn't willing to commit to them. Having a stake in their own future makes a huge difference. To move out of poverty, poor people have to invest their own time and money and they will do that if you offer them something that has a low level of risk while meeting their needs.

Polak takes the view that charity is appropriate for disaster relief but not for economic development (Fig. 3).

---

[7]Al-Hage (2009).

Fig. 3 Paul Polak, founder of International Development Enterprises, at the 2008 Pop! Tech conference. Photo by Kris Krüg. Detail from URL: https://flic.kr/p/5w9SUi

An important obstacle to the social entrepreneur approach to activism is that the marketplace often does not work in developing countries in the same way that it does in developed ones. For example, corruption is a frequent problem. Randy Schwemmin, technical director of D-REV (Design-Revolution, on which more below), notes that market failure occurred when his company sought to market a medical device called Brilliance in India. Brilliance was a machine that used blue light to treat severe jaundice in youth. It was engineered to cost only $400, as opposed to the $3500 alternatives from other sources. Even such a price differential was not enough to sell the machine[8]:

> But D-Rev realized early on that in India, the purchasing process wasn't working in Brilliance's favor. Hospital systems still sometimes chose higher-price systems because of bribery or cronyism, or because they didn't understand Brilliance's technical innovations, Mr. Schwemmin said.
>
> ...
>
> Plans to expand beyond India, meanwhile, hit serious bumps. One distributor in the Philippines ordered eight units from Phoenix for $500 each but then resold them for $2400, Mr. Schwemmin said. When D-Rev asked for the reason behind the drastic markup, the company said it needed to budget money for kickbacks, he said. Because of these and other experiences, "we feel the need to be a lot more involved in picking distributors and managing relationships, because we're afraid of corruption," he said.

In order for the machines to have their intended impact, D-REV had not only to carefully design its machine to function properly but also to be able to skirt corrupt distributors in developing markets. For example, parts for a product would have to be designed so that they could be manufactured by companies that D-REV knew were not open to corruption.

---

[8]Larson (2014).

In general, social entrepreneurism may be characterized as a social agenda as follows:

1. Vision: Economic opportunity for people in poverty.
2. Values: Affordability, productivity, marketability, …
3. Methods: Inexpensive materials, low maintenance, …

Case Study: The Remotion Knee

An instructive example of social entrepreneurism concerns the Remotion knee, also designed by D-REV. The Remotion knee is a prosthetic knee designed in cooperation with the Jaipur knee foundation, producers of the Jaipur foot described earlier.

The story of the development of the Remotion knee is given by Krista Donaldson, CEO of D-REV. Ms. Donaldson is from Halifax, Nova Scotia, and obtained her Ph.D. in (Mechanical) Engineering and Product Design at Stanford University. Her professional interests lie in applying engineering to people in poverty, which ultimately got her interested in the engineering of medical devices for people in developing countries (Fig. 4).

In her 2013 TED talk, Ms. Donaldson describes D-REV's approach to development of its products with special reference to the knee.[9] Watch to her presentation and identify the design principles that she describes.

Q: What ideas are central to Donaldson's approach?

Clearly, social entrepreneurism has many points in common with humanitarianism. There is great emphasis on affordability, collaborating with users, and scalability. However, commercial concerns also play a role. Ms. Donaldson mentions the concept of providing value for users. That is, the knee is designed to be viewed as a worthwhile investment for its adopters. Also, the entire manufacturing and distribution chain figures in design, and not only its impact on its intended audience.

Critique: Technological Colonialism

It may seem that design activism is self-evidently always a good thing. After all, who would object to increasing social justice in the world? However activism is not unquestionably appropriate on all occasions. There may be times when guerilla

---

[9]Donaldson (2013).

Fig. 4 Krista Donaldson,
CEO of D-Rev, presenting at
the 2011 Pop!Tech
conference. Photo by Kris
Krüg. Detail of URL: https://
flic.kr/p/axwqCj

design may be regarded as simple trouble-making, a view that airlines seem to take of the Knee Defender, for example.

Even humanitarianism may be questioned. Humanitarian design is often carried out by well-off designers from western countries who visit developing nations seeking to be helpful. Critics may not welcome such efforts, viewing them as a renewed form of colonialism, for example.

In the colonial era, military and commercial forces from developed nations took control or imposed themselves on peoples elsewhere in the world. The purpose of colonialism was typically exploitation of other nations and their resources for the benefit of the colonizers. Although this era of colonization is largely over, suspicion remains among many people in the developing world that developed nations still hold colonial attitudes towards them.

One such suspicion centers on the idea of _technological colonialism_. This term refers to suspicion that importation or imposition of western technology brings with it western agendas and experts, thus displacing local technology, agendas, and people. With western goods come western lifestyles and attitudes. By adopting cars and television sets, for example, the fear is that western car culture and media culture will be adopted with them, thrusting native culture aside.

Even humanitarian designers like Emily Pilloton, who emphasize respect for their clients, may viewed as colonialists, although perhaps inadvertently. Consider the following incident, reported by Bruce Nussbaum, a contributing editor for _BusinessWeek_ Magazine[10]:

> The last time I saw Emily was in Singapore in the fall at the ICSID World Design Congress where she was receiving a roaring applause from the European and American designers on stage after giving a speech about Project H. I loved that speech because it linked the power of design to the obligation to do good. In a world awash in consumption, with many designers complicit in designing that consumption, Emily's message was right on.

---

[10]Nussbaum (2010).

But not to the mostly Asian designer audience. Of course there was polite applause but, to my surprise, there was also a lot of loud grumbling against Emily along the lines of "What makes her think she can just come in and solve our problems?" This was a challenge of presumption that just stopped me cold—and sent me back to my Peace Corps days when I heard a lot about Western cultural imperialism from my Filipino friends. Are designers helping the "Little Brown Brothers?" Are designers the new anthropologists or missionaries, come to poke into village life, "understand" it and make it better—their "modern" way?

Then, some months later at Parsons School for Design, the same thing happened. I went to a talk by IDIOM Design, one of India's top design consultancies.

Might Indian, Brazilian and African designers have important design lessons to teach Western designers?

At the end of a great presentation, a 20-something woman from the Acumen Fund rushed to the front and said in the proudest, most optimistic, breathless way that Acumen was teaming up with IDEO and the Bill and Melinda Gates Foundation to design better ways of delivering safe drinking water to Indian villagers. She said this to the Indian businessman Kishoreji Biyani, who is the key investor in IDIOM, and to my stunned surprise—and hers —he groused that there was a better, Indian way of solving the problem. She didn't know what to say. And I didn't either.

I know the Acumen and IDEO people and they, like Emily, are the very best. I know the IDIOM folks and they, too, are the very best. And I have met Mr. Biyani in India and he is an amazing businessman. But he, too, like many in the Asian audience in Singapore, took offense at Western design intervention in his country.

Note that Mr. Biyani's objection to the proposed water system design was that it represented a western way of doing things rather than an Indian way. Even a humanitarian effort can be viewed as a kind of attack on local society. Of course, anything perceived as an attack will meet with resistance.

Such a response may seem puzzling. After all, if a design will increase social justice, then why not adopt it, no matter where it originated? To understand the situation better, it may help to imagine a reversal of roles. Suppose that designers from a developing nation visit a developed one and begin to make recommendations for improvement. For example, perhaps African designers could visit Canada and make recommendations for improving the lot of its indigenous peoples, many of whom live in poverty and without the kind of resources or medical care that other Canadians enjoy. Many Canadians would likely perceive such an intervention to be out-of-line, even if the suggestions were good ones.

Critique: Good Intentions

A related problem with activism from abroad is that activists may not see the real problems. There is a tendency for activists to show up, "understand" the situation, and then introduce a design fix. However, such quick fixes may do little to address issues of social injustice.

George Beane is a Peace Corps volunteer who performed humanitarian design work in Ecuador. Although a humanitarian designer himself, he expressed some misgivings about work he helped perform with the Pittsburgh chapter of Engineers Without Borders (EWB). The Pittsburgh crew repaired and upgraded a water pumping system for the village of Tingo Pucara. The project was successful in the sense that it provided flowing water to each tap in the village.

However, Beane worries that this success did not face the real problem[11]:

> The engineers I worked with were smart, selfless, dedicated people who gave up hours and money to work for people they typically never met. But they only ever saw part of the big picture, because the problem of getting water to poor people, treated as an engineering problem, is equally a problem of resource allocation, national policy, and local politics.
>
> …
>
> In the end, our massive expenditure of money and time helped a single village, about 120 people. Moreover, in the surrounding canton, village leadership is often split among families. When, as in the case of Tingo, one patriarch decides to relocate his clan on previously unoccupied land, the new settlement petitions regional governments and outside aid agencies for incorporation; and then for new infrastructure. Absent any regional planning, and with local mayors eager to please voting constituents, another water system is demanded. A different aid group steps into address the need.
>
> …
>
> It's easy to forget what bigger forces shape remote, impenetrable rural life. As I think about the results of our effort, I wonder if we didn't fall into that trap, if we didn't fail to look up and beyond the village horizon, and if maybe that wasn't the real problem from the start.

Q: How were EWB's efforts misplaced, according to Beane? Is he right?

Of course, the villagers in question were eager to receive a water system and it is only just that they should have access to one. However, Beane appears to feel that the constant demand for new water systems is a political maneuver by local patriarchs to maintain their social status within the community. Their ability to get water systems built supports their power. Thus, by designing and supplying such systems, activists are supporting a political regime that is, perhaps, unjust.

Of course, these designers have good intentions. However, good intentions may actually prevent them from seeing the agenda that their work is actually serving.[12]

These considerations indicate that designs intended to promote social justice are not necessarily good designs. What some see as social justice may strike others as vigilanteeism or colonialism. These concerns are important to keep in mind when assessing activism in design.

---

[11]Beane (2012).
[12]Shelley (2011).

References

Al-Hage, S. (2009, November 13). *Fighting against poverty*. Retrieved February 13, 2010, from Imprint: http://s3.amazonaws.com/UWPublications/Imprint/2009-10_v32/Imprint_2009-11-13_v32_i17.pdf

Beane, G. (2012, September 28). *Letter from Ecuador: Final assessment*. Retrieved September 30, 2012, from Metropolis Mag: http://www.metropolismag.com/Point-of-View/September-2012/Letter-from-Ecuador-Final-Assessment/index.php?tagID=562

Dean, A. O. (2002). *Rural Studio: Samuel Mockbee and an architecture of decency*. New York: Princeton Architectural Press.

Donaldson, K. (2013, December). *The $80 prosthetic knee that's changing lives*. Retrieved February 12, 2014, from TED: Ideas worth spreading: https://www.ted.com/talks/krista_donaldson_the_80_prosthetic_knee_that_s_changing_lives

Gadget Duck. (n.d.). *Knee Defender*. Retrieved June 02, 2013, from GadgetDuck.com: http://www.gadgetduck.com/goods/kneedefender.html

James Dyson Foundation. (2014). *MOM wins 2014 James Dyson Award*. Retrieved May 2, 2015, from James Dyson Foundation: http://www.jamesdysonaward.org/en-GB/news/mom-wins-2014-james-dyson-award-2/?cookies=true

Larson, C. (2014, January 12). *Light-bulb moments for a non-profit*. Retrieved January 14, 2014, from The New York Times: http://www.nytimes.com/2014/01/12/business/international/light-bulb-moments-for-a-nonprofit.html?_r=1

Nussbaum, B. (2010, July 6). *Is humanitarian design the new imperialism?* Retrieved July 10, 2010, from FastCompany: http://www.fastcodesign.com/1661859/is-humanitarian-design-the-new-imperialism

Pilloton, E. (2009). *Design revolution: 100 products that empower people*. New York: Metropolis Books.

PopTech. (2009, November 2). *PopTech 2009 Social Innovation Fellow Emily Pilloton*. Retrieved Oct 2, 2013, from Vimeo: https://vimeo.com/7393447

Rawsthorn, A. (2009, September 6). *Designing for humanity*. Retrieved October 7, 2013, from The New York Times: http://www.nytimes.com/2009/09/07/fashion/07iht-design7.html

Shelley, C. (2011). Motivation-biased design. *Proceedings of the 33rd Annual Conference of the Cognitive Science Society*, (pp. 2956–2961). Boston.

Social Spaces

Abstract Activism involves the view that good design may involve the pursuit of social justice. A significant domain for the application of social justice is in social spaces. Social spaces are places, public or private, where sizeable groups of people, who may not be known to each other, may meet or interact. So, the design domains of architecture and urban planning are central to this theme. Spatial justice is a concept that applies social justice to social spaces in particular. The concept of spatial justice derives from Henri Lefebrvre's concept of a right to the city. It concerns how people and the resources they need to thrive are distributed spatially, and how that distribution is decided on. Food desserts and gentrification are examples of problems that can arise where spatial justice is concerned. The phenomenon of urban activism is also discussed.

Introduction

The previous chapter concerned activism in design. This agenda involves design applied for improvement of social justice. Typically, this aim means changing the status quo in a society so that some obstacle to social justice is reduced or removed. Several approaches to activism were explored and some problems of this agenda examined.

In this chapter, we continue to examine activism in design but with focus on a particular sort of problem, that is, *social spaces*. A social space is a place where people routinely meet and interact, often people who are not known to each other and who appear in large numbers. Some social spaces are publically owned, like city parks. Others are privately owned, such as shopping malls. Whether public or private, a social space is where people encounter one another in substantial groups.

Social spaces are of special interest because they are fundamental to the social contract. Recall that the basic purpose of the social contract was to set some rules for people to follow so that they can enjoy their basic rights and deal with each other in mutually beneficial ways. Social spaces are where people encounter one another and, thus, their design is crucial to a working and healthy society.

C. Shelley, *Design and Society: Social Issues in Technological Design*,
Studies in Applied Philosophy, Epistemology and Rational Ethics 36,
DOI 10.1007/978-3-319-52515-0_9

The main concept that we will apply to evaluate social spaces is *spatial justice*. This concept concerns how the layout and furnishing of social spaces affords social justice, as well as how they are properly governed.

Case Study: Turn to the Future

The value of an apartment or condo in a tall building is determined in part by the view. Apartments that face water are more expensive than ones that face the opposite way. Apartments that are higher up are more expensive than ones lower down (or underground).

Before adoption of the elevator, the situation was reversed. Because the top floors of a building were hard to get to, the ground floor or the floor above were often the most valuable ones. Apartments near the ground floor were rented at the highest rates whereas rooms under the roof were reserved for artists (think of the artist's garret) and other people without much money to spend on rent.[1]

Today, elevators have removed the effort of reaching the top floors of tall buildings. Thus, buildings can be made much taller than before. Also, commanding views offered at the top means that apartments up there command higher prices.

Industrial designer Shin Kuo, of San Francisco's Academy of Art University, has proposed a conceptual design that would make all apartments in a building equal. In his proposal, a mechanical system would rotate individual apartments around a central core, so that they slowly descend down the exterior of the building before eventually reaching the ground floor. At that time, each apartment is raised to the top of the building to start its journey again. He calls this idea the Turn to the Future.[2]

Clearly, such a dynamic building would be more expensive to construct than a static one. However, Kuo argues that it provides a benefit to society that would make the expenditure worthwhile. Instead of having wealthy people permanently on top and monopolizing the views, every apartment dweller would have equal access to that amenity.

Q: Would the extra effort be worth the benefit?

Even if the design is never realized, the proposal reminds us that the distribution of living spaces in a building is not inevitable. The current distribution is a result of elevators and other technologies for constructing tall buildings. Innovations in building technology may someday prompt us to reconsider how buildings distribute access to amenities, such as scenic views.

---

[1]Bernard (2014).
[2]Metcalfe (2015).

Fig. 1 French philosopher Henri LeFebvre (1901–1991), originator of the concept of the right to the city. Photo courtesy of the Dutch National Archives, The Hague. URL: https://commons.wikimedia.org/wiki/File:Henri_Lefebvre_1971.jpg

⚹ Spatial Justice

The "revolving apartment building" situation illustrates an issue that is sometimes known as *spatial justice*. In brief, spatial justice refers to how social advantages and disadvantages are distributed throughout social spaces, especially urban areas.

The concept of spatial justice goes back to the writings of French philosopher Henri LeFebvre (Fig. 1).[3] With most of the world's people living in cities, the issue of how citizens should access and share city resources has become fundamental to civilization.

In its subsequent development, the right to the city has been parsed into two components[4]:

1. *Distributive*: The way in which resources are located can be considered just or unjust to different constituencies. Examples would include food deserts, walkability, public toilets, segregation;
2. *Procedural*: The way in which resource location is decided can be considered just or unjust. Examples would include gerrymandering, gentrification, redlining, fortification, NIMBYism, and marginalization.

The first point concerns how civic resources are distributed so as to be properly accessible to citizens. The second point concerns the right of citizens to determine collectively how to distribute those resources. We begin discussion of spatial justice with some examples of the issue of distribution.

---

[3]Lefebvre (1996).
[4]Cf. Soja (2010).

Case Study: Food Deserts

A good example of a distributive issue in spatial justice is the *food desert*. The concept of a food desert is basically simple: It refers to an area in a city where fresh food is not readily accessible. Fresh food would refer to things like produce and ground beef in distinction to processed or fast foods such as microwave burritos and ramen noodles. The concern is that, while people can survive on processed food, such foods are less healthy than fresh foods and therefore bad for the health of people who rely on them.

Consider the city of London, Ontario. A study conducted by Western University geographers Jason Gilliland and Kristian Larson found that London has food deserts, particularly in the east end of the city.[5] In their view, a food desert is an area where there is no supermarket selling fresh food within a 15-min walk (about 1 km) or a 10-min bus ride (without transfers).

Among other things, the researchers found that people in food deserts end up shopping at convenience stores, where food prices are 1.6 times higher than in supermarkets, on average. Also, food available in convenience stores is less nutritious, being higher in empty calories, for example. Not coincidently, people living in food deserts are at substantially higher risk of ailments such as heart disease, diabetes, and cancer.

In London, food deserts are located in areas of low-income residents who cannot afford cars to drive to grocery stores. Also, single mothers who cannot afford baby sitters to allow for long shopping trips are adversely affected, as well as people who face transport accessibility issues.

Q: How could food deserts be considered spatially unjust?

Given that good food is a necessity for thriving, it could be considered spatially unjust that people who are least able to afford it face special barriers in accessing it.

Case Study: A Right to Food

There are different solutions to the problem of food deserts in cities. Gilliland and Larson recommend that governments encourage grocery stores to open locations in such places. Certainly, that measure would help to alleviate the problem. However, not all food suppliers address the problem of food deserts equally well.

For example, Americans increasingly obtain their food from so-called big-box stores, that is, general goods stores housed in extremely large buildings, such as

---

[5]Larsen and Gilliland (2008).

Wal-Mart. In 2015, Wal-Mart announced that it had reached its objective of opening at least 275 stores within food deserts.[6] Although stores like Wal-Mart do carry fresh food, they also carry a great deal of processed food, which their customers tend to prefer. As Americans rely more heavily on such stores, the share of processed food in their diets increases at the expense of fresh food. Since processed food is higher in sugar, salt, and saturated fats, this trend suggests some people in these former food deserts will continue to suffer from poor nutrition.

A different approach has been applied in Brazil. In 1993, the city of Belo Horizonte declared that its residents have a *right to food*, that is, food deserts were made illegal.[7] To secure this right, the city set up programs such as subsidized public restaurants called *People's Restaurants* within food deserts. Also, they set up rolling markets called the *Big Basket* that bring food from neighboring farms into the city in specially modified busses. These busses have defined routes and schedules, allowing citizens along their routes to access good food cheaply, reliably, and conveniently. The effort has had some positive effects, such as increasing incomes for local farmers and lowering the infant mortality rate in the city.

These ideas have served as a model in other areas. In Chicago, for example, a non-profit group called *Fresh Moves* set up a rolling grocery store on a bus that drives through the low-rent areas of the city.[8] It makes regular stops, allowing locals to access the produce that it stocks.

Q: What other sorts of resource deserts might occur in cities?

Case Study: A Right to Pee

One sort of public resource that may be in short supply in social spaces is public washrooms. In Indian cities such as Mumbai, for example, there are very few such facilities for women.[9] There, it is considered acceptable for men to pee against nearly any walls or bushes. However, women may pee only in proper facilities, of which there are relatively few. Such facilities as exist are often non-functional.

The lack of washrooms poses significant problems for women who live in slums where there are even fewer private facilities:

---

[6]McMillan (2015).

[7]Lappe (2009).

[8]Lepeska (2011).

[9]Sachdev (2014).

Some women get bladder and urinary tract infections from holding in their urine, while others simply don't drink water all day to avoid the bathroom. Many women are raped or assaulted each year when they leave their homes to find a toilet, and those who find toilets safely can face other risks—scorpions, rats, infections.

Deepa Pawar, of the women's rights organization Vacha, argues that the lack of proper washrooms is the primary reason why young women drop out of school—in order to avoid these risks.

Where there are public toilets, women are often charged to pee there, unlike men who may use them for free. This practice further limits the ability of women, especially those low incomes, to move about freely in the public realm.

To address this situation, activists such as Pawar have formed the Right to Pee movement, advocating for more and safer public washrooms for women. Equal provision of public toilets would enable women to move about the city for purposes of work or provisioning in the same way that men are used to doing.

Q: Do women have a right to equal access to public washrooms?

Authorities in Indian cities tend to take the view that women belong in the home and should make limited use of public spaces in any event.

The unequal distribution of public washrooms in these locations can be considered as a kind of desert—a washroom desert. As such, it raises issues of distributional spatial justice.

Spatial Exclusion

Food deserts illustrate how citizens may lack access to some resource because it is located out of reach for them. The reverse can happen as well. That is, social spaces can be designed in order to exclude some people from using them even though they are close together. *Spatial exclusion* occurs when designs are made to keep a given constituency out of a social space.

A simple means of achieving spatial exclusion would be a wall. The Interboro partners, a firm of New York urban planners, discuss the example of a seawall in New Jersey.[10] State law requires public access to seafront beaches, which are officially considered public property. Access to these beaches is complicated by the presence of a seawall, in fact, an old railroad bed that forms a wall between the beaches and the interior of the state.

Homeowners next to the seawall form a powerful lobby group. They were allowed to build walkovers, i.e., bridges, to provide themselves with beach access from their own backyards. They were also successful in preventing any parking lots

---

[10]Armborst et al. (2013).

Fig. 2 Bench with armrests at a bus stop in New York City. Photo by Kristina Hoeppner. URL: https://www.flickr.com/photos/4nitsirk/9473839083/

from being built near the seawall, so that residents from other areas could not easily access the walkovers. In effect, these homeowners managed to monopolize the public beaches.

Under pressure of a lawsuit, the state Department of Environmental Protection (DEP) built a parking lot next to the wall but in a location where there were no walkovers available to the public. Following another lawsuit, the state DEP agreed to build four walkovers but sited them where there were no parking lots. In both cases, the public continued to be excluded from equal access to the beach.

A subtler example is a kind of armrest sometimes found on benches in public parks or bus stops (Fig. 2).[11] The benches are placed there to provide seating for members of the public. However, armrests built into the benches are designed not so much to provide a place to comfortably rest a sitter's elbows as to prevent anyone from lying down on them. Typically, the aim of these rests is to prevent homeless people from using the benches to sleep on. These armrests are an example of exclusionary design disguised as an amenity.

In both cases, walls and armrests are designed to keep a specific group from utilizing a social space in a given way.[12]

Q: What are some examples of spatial exclusion in your area?

---

[11]Armborst et al. (2011a).

[12]Cf. Savičić and Selena (2013).

Fig. 3 Angie and Carol
sitting on the stoop. Photo by
dianneb59@sbcglobal.
net/Flickr. URL: https://flic.
kr/p/5EvTtg

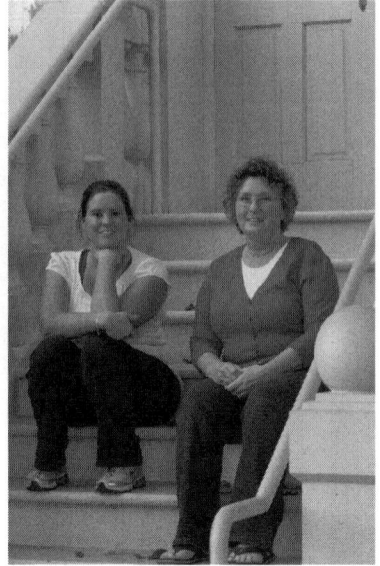

Fig. 3 Angie and Carol sitting on the stoop. Photo by dianneb59@sbcglobal. net/Flickr. URL: https://flic. kr/p/5EvTtg

Spatial Integration

The opposite of spatial exclusion is *spatial integration*, that is, designs that invite different constituencies to share a social space. The Interboro Partners provide some straightforward examples.

The first is the humble stoop.[13] A stoop is a short stairway that provides access to a private dwelling from a sidewalk. In many cities, people enjoy sitting out on their stoops, which provides them with opportunities to interact with passers-by and vice-versa. Through such usually peaceful interaction, local residents can establish a good rapport with their neighbors and even with people from outside their neighborhood (Fig. 3).

Another example is mass transit.[14] Busses, subways, and light rail systems allow people from different parts of a city to visit other parts without having to own or operate a car. People from different walks of life meet and interact within the mass transit system itself. Also, such systems allow broader access to common social spaces such as parks and shopping malls.

Of course, some areas of cities are designed to exclude mass transit. Suburban developments, for example, tend to provide roadways that are long and meandering. This design is navigable by car but difficult for mass transit. The amount of driving required to visit a large number of houses in a suburban settlement is usually uneconomical for a bus.

---

[13]Armborst et al. (2011b).

[14]Armborst et al. (2010).

Fig. 4 The Brooklyn Bridge.
Photo by Simone
Roda/Wikimedia commons.
URL: https://commons.
wikimedia.org/wiki/Category:
Brooklyn_Bridge#/media/
File:Brooklyn_Bridge_-_
New_York_City.jpg

Case Study: The Brooklyn Bridge

Sometimes spatial integration is an accidental side-effect of a design whereas, in other cases, it is a deliberate feature. In the accidental category, consider the Brooklyn Bridge (Fig. 4). This bridge was completed in 1883 by pioneering engineer John A. Roebling. It joined New York City on the island of Manhattan with the City of Brooklyn on the southern tip of Long Island.[15]

The original intention of the bridge was simply to facilitate commerce between the two cities. It certainly accomplished this goal. However, it also succeeded in uniting them. At the time, it seemed to some as though Brooklyn had a brighter future. Manhattan Island was beginning to fill up whereas Brooklyn had much more space to expand into. So, it seemed set to out-grow New York in size and also to build a larger port to out-compete it commercially.

However, the Brooklyn Bridge helped to establish a concept of commonality between the populations of the two cities. In 1898, the two cities joined together to form Greater New York.

Q: In what way did the Brooklyn Bridge increase spatial justice?

Though it was not the intention of its promoters, the Bridge certainly increased spatial integration, since it provided people on each side with better access to the

[15]Cf. Haw (2005).

people and opportunities that were available on the other side. It also helped to bring about the political integration of both social groups. Thus, residents of both areas got a say in the running of their common city, which increases the procedural aspect of spatial justice.

Case Study: Johannesburg

The lessons of the Brooklyn Bridge and others like it were not lost on urban designers. Today, urban bridges are sometimes designed with the express purpose of increasing spatial integration.

For example, in Johannesburg, South Africa, the city has been divided into mutually exclusive areas that have been described as a kind of "spatial apartheid", with well-to-do white people living in the city's wealthiest suburb, Sandton, and low-income, black people living in Alexandra Township (where Nelson Mandela once lived). The city's biggest highway separates the two, making it difficult for the 10,000 or so residents of Alexandra who work in Sandton to commute.

To mitigate the problem, Johannesburg is building a pedestrian and bicycle bridge over the highway to connect the two neighborhoods. The result should help commuters and also promote integration between the two sides[16]:

> In 2013 the executive mayor introduced the Corridors of Freedom as areas where there can be walking, cycling and public transport which is safe, reliable and affordable. This bridge satisfies that basic need that talks to citizens' rights to a spatially integrated city.

We tend to think of bridges in strictly technical terms, that is, in terms of the traffic that they can carry. However, urban bridges link communities, potentially affecting spatial justice as a result.

Q: What are some examples of spatial integration in your area?

Gentrification

So far, we have focused on the distributive aspect of spatial justice. That is, we have looked at how people are located relative to resources they may need access to. As noted above, spatial distribution also has a procedural aspect. This aspect of spatial justice relates to how spatial distribution is determined.

---

[16]South Africa Info (2014).

Of course, a powerful way in which the layout of cities is determined is the marketplace. In other words, structures and spaces go where people can afford to put them, and where they will deliver profits. Although this system has its advantages, advocates of spatial justice point out that it can create problems as well. One of those problems goes by the name of gentrification.

Gentrification is an increasing concern in spatial justice. Broadly speaking, gentrification refers to what happens to a low-income, urban neighborhood when higher-income people begin to move in.[17] Although there is no canonical definition of gentrification, it tends to involve the following changes[18]:

1. *Demographic*: An increase in median income, a decline in the proportion of racial minorities, and a reduction in household size, as low-income families are replaced by young singles and couples.
2. *Real Estate*: Large increases in rents and home prices, increases in the number of evictions, conversion of rental units to ownership (condos) and new development of luxury housing.
3. *Land Use*: A decline in industrial uses, an increase in office or multimedia uses, the development of live-work "lofts" and high-end housing, retail, and restaurants.
4. *Culture and Character*: New ideas about what is desirable and attractive, including standards (either informal or legal) for architecture, landscaping, public behavior, noise, and what constitutes a public nuisance.

Gentrification brings some significant benefits. For example, gentrified areas often see a reduction in crime rates. In addition, they tend to experience an upswing in economic activity and property taxes.

At the same time, gentrification brings some significant problems. For example, gentrified areas often sustain a loss of industrial employment, such as factory work. In addition, the lower-income residents are often marginalized or displaced to less desirable areas.

Gentrification is a problem of social justice because these advantages are realized mostly by the incoming residents and the disadvantages are realized mostly by the displaced ones.

Gentrification can be considered a procedural, spatial injustice in the sense that the people who are displaced often have no recourse, no way to appeal, as it were, their treatment. They can complain to city governments but, historically, those bodies are most attentive to higher-income residents. Furthermore, governments sometimes favor gentrification because it increases property values and thus property tax revenues. So, governments have incentives not to heed complaints on the matter.

---

[17]Cf. Lees et al. (2008).
[18]Grant.

Fig. 5 High Line Park, NYC. Photo by David Berkowitz. URL: https://flic.kr/p/a2rCko

Case Study: High Line Park

Most people like parks. Parks provide recreational opportunities for adults and children. They provide green spaces that people find relaxing or therapeutic. They provide places where neighbors can meet and people can get to know and trust one another.

However, parks can also contribute to gentrification. Locating a park in a given neighborhood tends to attract higher-income occupants, thus giving rise to the changes noted above.

This process has occurred recently in New York City, with the construction of the High Line Park (Fig. 5). In 2006, the city began to rebuild a 2.3 km disused section of railroad track of the New York Central Railroad. This was an elevated railroad formerly used by commuters in the city. Basically, the tracks were ripped out and replaced by pathways and gardens, accessible by ramps and stairs. Now people can enjoy walking through the city in a pleasant landscape with attractive views. Naturally, they like it.

However, the presence of the Park has also attracted higher-income people and new investments in surrounding infrastructure. As is often the case, this migration has resulted in changes such as the conversion of old factories into condominiums and the appearance of more high-end services such as Starbucks. Consequently, property values and taxes have increased by 103%, and working class people and jobs have begun to be displaced.[19]

Q: How could construction of a university lead to gentrification?

[19]Jaffe (2014).

Case Study: Pop-up Parks

One sort of response to the problem of gentrification is for city governments to intervene in the function of the marketplace. City governments may be able to provide support for less well-off neighborhoods through initiatives such as affordable housing.

However, instead of top-down solutions coming from government, advocates of spatial justice favor bottom-up approaches where residents of a neighborhood have the initiative in the design of their social spaces.

One example would be a *pop-up park* (Fig. 6). A pop-up park is a small, self-contained space that can be established in a closed-off area and provide the residents (and anyone else) with some amenity that is otherwise missing. Pop-up parks may contain just a few benches where people may sit and converse, or bike racks and play equipment for more active and large-scale recreation.

In Los Angeles, a city initiative called People Street allows neighborhood groups to apply for and receive a pop-up park kit. The kits come in different configurations called parklets.[20]

> For parklets, the initial step requires choosing a model. A parklet can be arranged as a classic café, a landscaped lounge with sloped stadium seating, or a sidewalk extension with simple seating and planters. Each parklet model is modular, offering applicants three different choices of design, or nine total to choose from. The furnishings are mostly movable, so they can be arranged for either small or larger groups.

> Once a model is chosen, community partners must select a color scheme for decking, furnishings, painted perimeters and roadbed graphics. The roadbed graphics are typically brightly colored, striped or polka-dotted. The kits also outline required safety features like planters and reflective border posts to make drivers more aware and encourage greater caution.

Pop-up parks can address the distributional problem of a lack of park space in a given neighborhood. In addition, they are selected and configured by people in the neighborhood where they are to appear, which represents a bottom-up approach to park allocation.

Q: Are pop-up parks a good solution?

Pop-up parks are an increasing popular feature of urban design.

---

[20]Gluck (2014).

Fig. 6 Using the pop-up park. Photo by Hrag Vartanian. URL: https://www.flickr.com/photos/hragvartanian/3931503003/

Urban Activism

Pop-up parks represent a bottom-up procedure for enhancing spatial justice to the extent that members of a neighborhood have the ability to apply for them from the city government. However, there are times when city governments are not so open to suggestion. In such cases, activist designers may resort to more aggressive, bottom-up procedures.

Urban activism may be characterized as intervention in a social space for the purpose of protesting and perhaps correcting action or inaction on the part of civic authorities.

For example, in 1997, the city of Los Angeles decided to erect a fence around a small neighborhood park in Santa Monica known as Triangle Park. It had become something of a hangout for homeless people and also sometimes for muggers. After police and neighborhood residents got fed up with these difficulties, the city erected the fence around the park, a fence that keeps everyone out (with the exception of the Department of Public Works employees for occasional groundskeeping).

An anonymous group of urban activists calling themselves "Heavy Trash" decided to protest this action by building a bridge over the fence. The bridge provided stairs so that residents who still wanted to access the park could do so. The city was evidently not amused and removed the stairs after three weeks.[21]

---

[21]Cf. Richards (2005) and Heavy Trash (2005).

Case Study: Tactical Urbanism

Tactical urbanism is an interesting example of urban activism. It is typically involves a group of designers who help residents to improvise new social space designs for themselves after they have failed to get city governments to act on their behalf.

For example, Mike Lydon of *Street Plans Collaborative* visited the city of Hamilton at the invitation of the *Hamilton-Burlington Society of Architects* to advise them on how some of the city's more problematic intersections might be redesigned (Fig. 7). Specifically, problems related to a school zone where drivers tended to drive too fast down a wide, one-way street and then not stop properly at the nearby intersection. Residents had complained to city officials, who did not take any immediate action.

Lydon recommended some actions, including having citizens paint a crosswalk where none was placed officially, and using traffic cones to narrow the street at a crossing point near the school. The bump out won the approval of the crossing guard who worked at the scene[22]:

> I asked the long-time crossing guard what she thought of the project. With immediate enthusiasm, she said, "I like it!" The guard did not know who had installed the cones or why, but she was highly supportive, saying it makes the corner a lot safer.

> The traffic calming "really controls the traffic. It was getting scary," she said, noting that the bumpouts force the cars to slow down instead of racing aggressively through the intersection.

However, the City of Hamilton was not thrilled with the changes and said so in an official letter to the public:

> These changes to City streets are illegal, potentially unsafe and adding to the City's costs of maintenance and repair. The City can consider this as vandalism, with the potential for serious health and safety consequences for citizens, particularly pedestrians. There is potential liability and risk management claims to both the City and the individuals involved.

The city removed the emendations. However, after meetings with concerned citizens, the city agreed to establish official crosswalks and bump outs at the controversial sites.

Q: Did tactical urbanists do the right thing in this case?

Other examples of tactical urbanism would include improvised street furniture, such as seats and benches, and "guerilla gardening", such as the planting of flowers or vegetables in verges and medians.

---

[22]Goodyear (2013).

Fig. 7 Mike Lydon, author
of *Tactical urbanism* (2015).
Photo by Aimee Custis.
(Lydon and Garcia 2015)
Detail of URL: https://flic.kr/
p/sMKvb3

Problems of design assessment of social spaces can be considered in at least two ways. First, Lefebvre's concept of the right to the city can be considered as a way of applying the concept of social contracts to social spaces. It concerns rights that people may have regarding social spaces, and how can social spaces' designs respect or disrespect those rights.

Second, the concept of *spatial justice* can be applied to social spaces as a special kind of social justice. When considered in this way, the possibility of activism arises in connection with how social spaces are designed, and by whom.

References

Armborst, T., D'Oca, D., & Theodore, G. (2010, September 5). *Light rail.* Retrieved October 17, 2013, from Arsenal of exclusion and inclusion: http://arsenalofexclusion.blogspot.ca/2010/09/light-rail.html

Armborst, T., D'Oca, D., & Theodore, G. (2011, November 29). *Arsenal of exclusion and inclusion.* Retrieved October 17, 2013, from Armrest: http://arsenalofexclusion.blogspot.ca/2011/11/armrest.html

Armborst, T., D'Oca, D., & Theodore, G. (2011, September 1). *Stoop.* Retrieved October 17, 2013, from Arsenal of exclusion and inclusion: http://arsenalofexclusion.blogspot.ca/2011/09/stoop_3000.html

Armborst, T., D'Oca, D., & Theodore, G. (2013, September 25). *Sea Bright and the sea wall.* Retrieved October 17, 2013, from The arsenal of exclusion and inclusion: http://arsenalofexclusion.blogspot.ca/2013/09/sea-bright-and-sea-wall.html

Bernard, A. (2014). *Lifted: A cultural history of the elevator.* (D. Dollenmayer, Trans.) New York: New York University Press.

Gluck, M. (2014, April 24). *How L.A. designed simple kits that let you 'make-your-own' park.* Retrieved April 25, 2014, from FastCompany: http://www.citylab.com/politics/2014/04/how-l-designed-simple-kits-let-you-make-your-own-park/8689/

Goodyear, S. (2013, June 5). *Painting your own crosswalk: Crime or civic opportunity?* Retrieved June 8, 2013, from The Atlantic: http://www.citylab.com/commute/2013/06/painting-your-own-crosswalk-crime-or-civic-opportunity/5791/

Grant, B. (n.d.). *What is gentrification?* Retrieved October 17, 2015, from PBS.org: http://www.pbs.org/pov/flagwars/what-is-gentrification/

Haw, R. (2005). *The Brooklyn Bridge: A cultural history.* New Brunswick, NJ: Rutgers University Press.

Heavy Trash. (2005, June). *Stair to park.* Retrieved May 12, 2008, from Heavy Trash: http://heavytrash.blogspot.ca/2005/04/stair-to-park.html

Jaffe, E. (2014, October 15). *How parks gentrify neighborhoods, and how to stop it.* Retrieved October 17, 2014, from FastCompany: http://www.fastcodesign.com/3037135/evidence/how-parks-gentrify-neighborhoods-and-how-to-stop-it

Lappe, F. M. (2009, February 13). *The city that ended hunger.* Retrieved October 12, 2013, from Yes! Magazine: http://www.yesmagazine.org/issues/food-for-everyone/the-city-that-ended-hunger

Larsen, K., & Gilliland, J. (2008). Mapping the evolution of 'food deserts' in a Canadian city: Supermarket accessibility in London, Ontario, 1961–2005. *International Journal of Health Geographics, 7*(1).

Lees, L., Slater, T., & Wyly, E. K. (2008). *Gentrification.* New York: Routledge/Taylor & Francis.

Lefebvre, H. (1996). The right to the city. In *Writings on cities* (E. Kofman, & E. Lebas, Trans.). Cambridge, MA: Wiley-Blackwell.

Lepeska, D. (2011, November 18). *Grocery stores on wheels.* Retrieved November 22, 2011, from The Atlantic: http://www.citylab.com/commute/2011/11/grocery-stores-wheels/528/

Lydon, M., & Garcia, A. (2015). *Tactical urbanism: Short-term action for long-term change.* Washington, D.C.: Island Press.

McMillan, T. (2015, October 13). *Why Wal-Mart and other retail chains may not fix the food deserts.* Retrieved October 15, 2015, from National Public Radio: http://www.npr.org/sections/thesalt/2015/10/13/448300139/why-wal-mart-and-other-retail-chains-may-not-fix-the-food-deserts

Metcalfe, J. (2015, April 17). *Everyone has a great view in a spinning apartment building.* Retrieved April 20, 2015, from The Atlantic: http://www.citylab.com/housing/2015/04/everyone-has-a-great-view-in-a-spinning-apartment-building/390768/

Richards, A. (2005, July). Gate crashers. *Los Angeles,* pp. 24–26.

Sachdev, C. (2014, November 25). *Women in India agitate for their right to pee.* Retrieved March 12, 2016, from Public Radio International: http://www.pri.org/stories/2014-11-25/women-india-agitate-their-right-pee

Savičić, G., & Selena, S. (Eds.). (2013). *Unpleasant design.* Belgrade: Akademija.

Soja, E. W. (2010). *Seeking spatial justice.* Minneapolis, MN: University of Minnesota Press.

South Africa Info. (2014, June 14). *'Iconic' walkway bridge to connect Alexandra, Sandton.* Retrieved October 12, 2014, from SouthAfrica.info: http://www.southafrica.info/business/economy/development/bridge-050614.htm

Risk = danger

Abstract In terms of social contracts, designs may be assessed according to how well they respect people's rights and how well they promote social justice. One limitation of this approach is that it omits the importance of prediction in design assessment. That is, designs are configured according to assumptions about what the future will be like. As noted in our discussion of rational design, predictions about the future can be inaccurate. Yet, our discussion of designs and social contracts has taken no account of this fact. Fortunately, there are ways of assessing designs in light of uncertainty about how the future will turn out. In this chapter, the concept of *risk* is introduced. Risk assessment refers to the analysis of uncertain future impacts of decisions and is readily applied to design assessment. The expected-value model of risk is described and applied to several cases of design assessment. On this model, designs may be assessed by scrutinizing the distributions of risk that they may give rise to. In particular, the principles of *collectivism*, *equity*, and *individualism* in the distribution of risk are examined.

Introduction

In the previous few chapters, we have looked at design assessment through the lens of the social contract. In essence, this perspective involves envisioning how the world should be arranged so that people can thrive and then designing things to help bring that vision about.

One issue that arises with this approach is that it is not always clear how designs will affect people. From our discussion of rationality and unintended consequences, we know that the future will bring surprises, that is, things will not always turn out as they are meant to. In short, the future is uncertain.

For this reason, uncertainty is an important issue for assessment of designs. How can we make sensible, moral assessments of the future impact of designs if it is uncertain how they will affect people? One answer to this question is to frame assessments in ways that take account of that uncertainty.

© Springer International Publishing AG 2017
C. Shelley, *Design and Society: Social Issues in Technological Design*,
Studies in Applied Philosophy, Epistemology and Rational Ethics 36,
DOI 10.1007/978-3-319-52515-0_10

One concept that has been developed for this purpose is *risk*. Roughly speaking, risk means the same as "danger", that is, a characteristic of situations that may involve significant harm. Many designs are intended to operate in situations where significant harm to people could occur. As a result, risk is a useful tool for their evaluation.

In this chapter, the concept of risk is considered and a common model of risk introduced. We then examine several principles that describe how risk ought to be distributed to people through the performance of designs that they use. These principles are *collectivism, equity,* and *individualism*.[1] These principles allow us to assess designs by determining how they mete out risk and then questioning whether or not that distribution is appropriate.

Case Study: Semi-automated Cars

Design scholar Donald Norman (Fig. 1) has long criticized the strategy of partial automation.[2] For example, there are cars on the market that have partial automation systems such as automatic lane following and automatic cruise control. Such cars can drive themselves for extended periods, at least on highways where the roadway is fairly simple and cars mostly stay in one lane for long periods. Only when an unusual situation arises are the automatic systems unable to drive the car properly, e.g., in view of sudden lane changes by other cars. At such times, the automatic systems often shut off and the driver must take control of the car.

Norman previously argued that partial automation is a mistake. It tends to encourage inattention in drivers. After all, the job of supervising an automated system is decidedly boring. As a result, drivers soon cease to pay much attention to the road and are then not in a position to drive properly in a sudden emergency. For example, the driver of a Tesla model S—a car equipped with an "autopilot" mode—was recorded apparently sleeping behind the wheel while his car was driving itself.[3]

However, Norman has recently changed his mind.[4] Since drivers' attention is now routinely distracted by items such as smart phones, the risk of bad driving from distraction is overtaking the risk of bad driving through inattention. In order to reduce the overall risk of crashes on the roadways, partial automation is better than none, he says. In other words, the risk of crashes through inattention is a worthwhile trade-off against the risk of crashes through distraction.

Other responses are possible. For example, drivers with smart phones could be required to use a "driving mode" that defers distractions until they can be dealt with

---

[1]These terms are adapted from Hansson (2013).

[2]Norman (1990).

[3]Whitten (2016).

[4]Norman (2015).

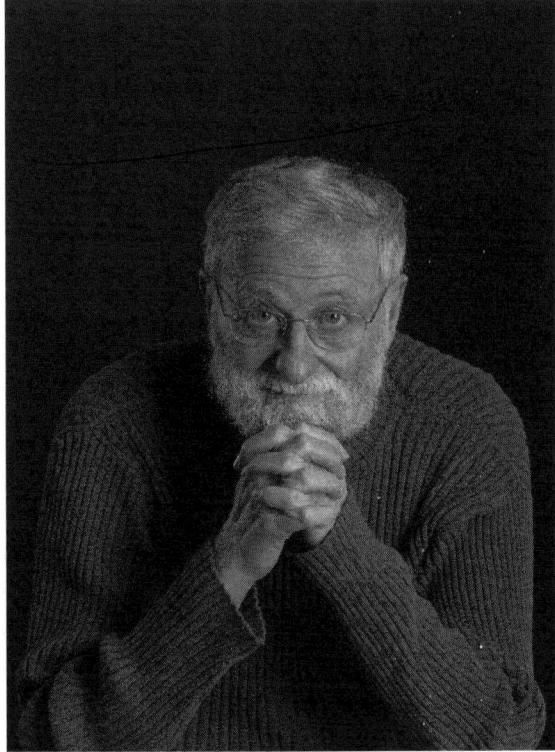

Fig. 1 Donald Norman, design scholar and author of *The design of everyday things* (2013). URL: http://www.jnd.org/NNg-Photographs/belanger/photo_5.jpg

safely.[5] This view might be justified by the argument that since distracted drivers are the source of the unusual risk, then they should be the focus of its solution.

Q: Which solution would be better, partial automation or reduction of distractions?

Risks

In everyday English, the term *risk* is roughly equivalent to *danger*. That is, when we take a risk, that means we are exposed to some potential for harm. Even if the harm does not befall us, the risk was there. For example, if a person crosses a street,

---

[5]Cf. The Economist (2013).

then there is a danger of being hit by a car. Even if there is no incident, the danger was still present.

This observation suggests that risk involves uncertainty. Even when someone faces a risk, the associated harm may not materialize. It is important then, that our notion of risk should reflect the fact that risks are present even when harms do not occur.

In addition, risks are often compared to one another. That is, we often judge one risk to be greater or lesser than a second one. For example, habitual smokers run a greater risk of lung cancer than do non-smokers, other things being equal. Furthermore, the more a person smokes, the more likely that person risks lung cancer.

The fact that risks can be compared leads people to make risk trade-offs. That is, they find themselves in situations where they face one risk or another and have to decide which one they prefer to take. The trade-off between inattention and distraction is central to Donald Norman's argument in favor of partial automation of cars, for example.

Let us return to risks involved in different methods of crossing a city street. In some cases, there will be several ways to make a crossing. A common one would be signalized crosswalk at a street corner, where pedestrian crossing is synchronized with traffic. Another possibility in some cases would be a pedestrian overpass or a tunnel beneath the street, perhaps built as part of a subway system.

Q: What are some risks of each crossing method? Why would you prefer one method to the other?

Risks of using a crosswalk include being hit by a car, perhaps driven by a distracted driver, and slipping on the roadway, perhaps due to bad weather. Risks of using a bridge or underpass include falling on stairs while getting in or out, being delayed due to the extra time it takes to use the bridge or tunnel, and being assaulted in the confined space while being unable to run away or call for aid.

A Formal Model of Risk

A concept of risk should represent both its element of uncertainty and the comparability of risks, so that trade-offs may be considered. The following model of risk is often used for this purpose[6]:

$$\text{Risk of event} = (\text{Probability of event}) \times (\text{Severity of event})$$

---

[6]Cf. Aven (2012).

The probability of an event may be the frequency with which it occurs. This factor represents the uncertainty of risk. The severity of an event represents the harm or cost caused in the event of its occurrence. This factor allows risks to be compared to one another.

For example, the average risk of having a fall by traversing a flight of stairs that requires hospitalization would be the product of the probability of such a fall (1/3,616,667 trips up or down a flight of stairs)[7] times the severity of the fall (ca. $5567 CDN average cost per hospital stay in Canada)[8]:

$$1/3, 616, 667 \times \$5567 = \$0.00154$$

The risk of this event is small, in part, because the probability of its occurrence is small.

Comparability of risks can be achieved if common units of measure for severity are used. In this case, the severity of hospitalization has been represented in terms of its financial cost.

Severity is often represented in terms of prices because prices do a good job of representing the cost of goods and services that are broadly available in the marketplace. The price of a good represents what people are willing to pay to get it and what value is lost if the good is taken away.

Of course, prices may do a poor job of representing the severity of non-market goods. For example, a life is a non-market good—one cannot be bought on Ebay, for example—and so has no market value. This fact makes risks of loss of life or quality of life harder to represent and compare. However, that is a complication that we will set aside here.

Safety Trade-Offs

The point of using risk in design assessment is to understand and evaluate how designs distribute risks among people. Oftentimes, designs reduce risks for one part of the population but increase it for another part. That is, safety is often a trade-off.

Consider the use of red-light cameras at intersections (Fig. 2).[9] In terms of safety, the purpose of the cameras is to discourage—by threat of fines—drivers from driving through intersections against red lights. The safety benefit of these

---

[7]Bryson (2010), p. 309.
[8]Kusch (2014).
[9]Huang et al. (2006).

Fig. 2 Traffic camera in
Tallahassee, Florida. Photo by
Michael Rivera/Wikimedia
commons. URL: https://
commons.wikimedia.org/
wiki/Category:Red_light_
cameras#/media/File:Traffic_
camera,_US319,_Tallahassee.
JPG

cameras is that they reduce the probability of side-on (or "t-bone") collisions. Side-on collisions may occur when a car running a red light crashes into a car crossing the intersection with the green light from a right angle. Such collisions are quite harmful, especially to the people whose car that is hit on its side.

However, red-light cameras can increase rear-end collisions. The reason seems to be that drivers who approach an intersection and see an amber light are likely to brake suddenly to avoid getting photographed and fined. Drivers following behind who are too close or not paying attention may then hit the car that is stopping quickly in front of them. Yet, rear-end collisions are usually less harmful than side-on collisions because a car's engine and trunk compartments are able to absorb a substantial amount of such impacts.

As a result, red-camera lights generally achieve an overall increase in safety, that is, an overall decrease in risk. This overall result is achieved in part by trading off one sort of collision for another sort.

> Q: What other safety features involve risk trade-offs?

Another example might be the modern office chair. Modern office chairs are often highly adjustable in back support, arm height and angle, seat pan depth, and provide comfortable fabrics and cushioning. Through increased ergonomics, office chairs have decreased risks of spinal injury or simple discomfort. At the same time, they may have increased injuries incurred from prolonged sitting, such as thrombosis, or from lack of exercise.[10]

---

[10]Tenner (1997), pp. 168ff.

Collectivism — *serves the common good.*
- *universal/impartial*
- *Requires that some people pay a price for the good of others*

Safety features in designs sometimes trade off one form of risk for another one. Designers then face the question: When is a safety trade-off acceptable? Sven Ove Hansson is a Swedish philosopher who has identified several principles of risk distribution that might be used to answer this question (Fig. 3).

One answer is that a trade-off is acceptable when it provides the greatest overall benefit against the alternatives. Hansson calls this view the collectivist principle.

> Collectivism: An option is acceptable to the extent that the sum of all individual risks that it gives rise to is outweighed by the sum of all individual benefits that it gives rise to.

On this principle, the best design is the one that leads to the best—that is, safest—average outcome amongst all alternative designs. The risk posed to every individual by a design is calculated, and the total risk to all individuals added up. This calculation is made for every design alternative and the best one is then identified as the one with the best score. The process is identical to Herbert Simon's characterization of optimal design, with minimal, overall risk as the utility function.

Moral justification for collectivism in risk assessment comes from appeal to the common good. The common good refers to a situation that is the best for everyone concerned, where the interests of every individual or group are regarded as equally important. Thus, the common good exhibits two qualities that make it morally appropriate, namely universalism and impartiality. It is universal in the sense that all risks brought about by a design decision are considered in this assessment. It is impartial in the sense that all risks are given equal weight; no one gets more consideration than anyone else.

For example, proponents of red-light cameras argue that having these cameras is better than not having them because the overall benefit of the collisions avoided by having the cameras outweighs the risks of collisions that result from having the cameras. All risks are considered and no one's risks are given special consideration.

One of the important implications of collectivism is that some people may well be harmed in order for others to escape injury. In the case of red-light cameras, some people will experience (sometimes fatal) rear-end collisions when they otherwise would not in the absence of those cameras. This observation identifies an important objection to collectivism, that it may involve sacrificing the good of some people in order to achieve good for others.

Externalities

One of the virtues of collectivism is that it requires us to consider risks to everybody, leaving nobody out. The importance of this consideration may be illustrated in cases where significant risks are indeed left out. Designs that receive a positive review only because significant risks are not accounted for are suspect.

Fig. 3 Sven Ove Hansson,
scholar of risk and author of
The ethics of risk (2013).
Photo by Marco Blomberg

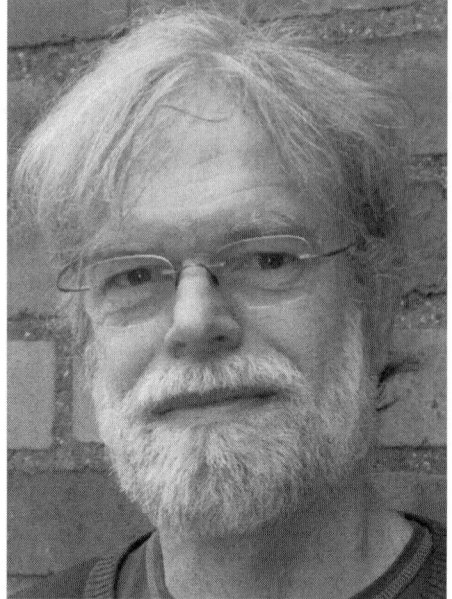

Consider the case of air conditioning for large apartment buildings. Very often, such buildings receive individual air conditioners, one for each unit, as noted by John Moyers (an environmental consultant) about the design of many apartment blocks:[11]

> ...he cites the HVAC (heating, ventilating, and air conditioning) systems used in many of the upscale housing units that have been built by Donald Trump and others in New York City. The cheaper (to install but not to operate) under-window, through-the-wall units pass costs to the tenants and noise—'this dull roar'—to the street. Of course, one common way that urban dwellers attempt to mask street noise is to turn on *their* air conditioners.

Apartments in these large blocks are designed to use individual air conditioners because that design is less expensive for the developer. However, individual units impose costs on subsequent apartment dwellers and their neighbors in terms of noise. Central air-conditioning, although more expensive to install, would greatly reduce the problem of noise, both for apartment dwellers and their neighbors.

Noise pollution, such as that generated by a chorus of air conditioning units, is associated with health risks. Besides hearing loss, there is hypertension (high blood pressure), heart disease, and disturbed sleep.

This case is an example of an *externality*, that is, a risk that results from a design that is imposed not on its producers or purchasers but on third parties. In other words, the risk is external to the arrangement that brought it about. In this case,

---

[11]Keizer (2010), p. 61.

Fig. 4 A graphical representation of an externality. Here, the public suffers risks that go unconsidered by designers and their clients

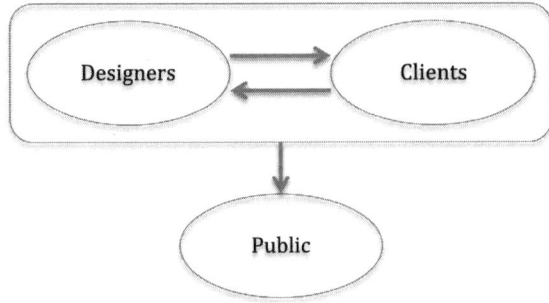

risks resulting from noise pollution generated by individual air conditioners is suffered by apartment dwellers and neighbors and not designers or developers.

See Fig. 4 for a graphical representation of an externality. In it, risks imposed on the public are represented as being external to the relationship between clients and designers, and thus not accounted for in their evaluation of a design.

Collectivism in design evaluation requires elimination of externalities because everyone's interests must be considered. It would arguably lead to a better result in the case of apartment buildings and their air conditioning arrangements.

Q: What other design externalities can you think of?

Many forms of pollution qualify as externalities. If a design gives off something that can be considered pollution, then the collectivist principle suggests that it is not a good design.

So, collectivism in design assessment may be justified on the basis that it helps to eliminate unacceptable externalities.

Equity

One of the difficulties with collectivism as a way of distributing risk is that it may involve decreasing the welfare of some people in order to achieve a better result for the greater society. Using—or abusing—one group of people to help others can be problematic.

A different principle of risk distribution that addresses this issue is *equity*. There are different treatments of the concept, but it may be framed in this way:

> *Equity*: An option is acceptable to the extent that it is part of a social system of risk-taking that is mutually advantageous and in which all participants enjoy equal latitude and consideration. (Hansson)

Here, *latitude* refers to the range of things that participants are allowed to do if they choose. *Consideration* refers to allowances that others ought to make for them. For the purpose of illustration, consider the case of automotive mobility:[12]

> To make this more concrete: as car-drivers we put each other's lives at risk. However, if we are all allowed to drive a car, exposing each other to certain risks, then we can all lead more mobile lives, and this will on balance be to the benefit of all of us (or so we may assume).

Let us consider each aspect of this case in turn. First, allowing everyone the opportunity to drive a car is mutually advantageous. After all, mobility is crucial to modern living, such as commuting to and from work.

Also, every potential car-driver has the same latitude in the sense that they all have permission to put every other car-driver at some risk of injury or death. Furthermore, all car-drivers are given the same consideration in the sense that each driver is exposed to the same kind of risks as all the other drivers. In addition, this arrangement places limits on those risks. For example, risks posed by drunk driving or distracted driving are not included. Each driver is supposed to be free from risks of that nature. So, people who drive while drunk or while distracted are operating outside of this arrangement.

The idea of equity is best understood through the concept of the social contract. Hansson argues that everybody has a basic right to personal safety, that is, a right to be free of risk imposed upon them by others. However, people may compromise on this right in order to gain access to some mutual benefit. In the case of automotive mobility, people compromise on their right to personal safety by allowing others to place them at risk through driving. They do so in order to gain access to benefits of driving cars themselves. This compromise is justified because it puts drivers in a better position to thrive, which is the basic goal of the social contract. The social contract that applies to driving then specifies which risks are acceptable (e.g., rain or fog) and which ones are not (e.g., drunk driving).

As Hansson points out, the equity principle differs fundamentally from the collectivist principle. On the collectivist principle, people are used simply as a means of distributing risk to achieve an overall result. No consideration of anyone's rights is given. On the equity principle, people's rights to safety are fundamental. A good design, then, is one that distributes risk equitably.

Case Study: Cyclist Alert

Roadways are shared by many types of vehicles. As noted above, this situation is equitable so long as all users of the roadways are at roughly equal risk. However, although bicycles are considered vehicles for the purposes of most roadways, cyclists are at substantially higher risk than are motorists.

---

[12]Hansson (2013), p. 1097.

Roadways themselves are designed primarily with the safety of motorists in mind. Also, although collisions with cars pose a particular threat to cyclists (and not often the reverse), safety features on cars have not often taken the safety of cyclists into account. This lack of consideration makes road safety inequitable for cyclists.

This situation is not inevitable and road and car design have begun to pay greater attention to it. Consider an in-car warning system being developed by Jaguar to alert drivers to the presence of cyclists who may be danger.[13] The system includes sensors that detect cyclists and pedestrians in the driver's blind spots and warns drivers either through taps on their appropriate limb or through warning sounds, e.g., a bike bell sound on the appropriate side of the car. The system would also prioritize warnings so that they do not become confusing or overwhelming.

By increasing consideration given to risks posed to cyclists by cars, Jaguar's "Bike Sense" system would improve equity of safety among vehicles on the roadway.

> Q: What other examples of safety equipment increase equity?

The proposed "Bike Sense" system illustrates how considerations of equity can appear in design assessment. A design that tends to place members of a given social group at significant disadvantage is inequitable and therefore not a good design. Such situations may be addressed through inclusion of safety features that limit risks to those disadvantaged people.

Individualism

Both collectivism and equity emphasize how to assess the appropriateness of risk trade-offs across groups of people. However, there are occasions where the social distribution of risk is not relevant to design assessment. In such cases, the principle of individualism may apply:

> *Individualism*: An option is acceptable to the extent that the risk to which each individual is exposed is outweighed by benefits for that same individual.

In other words, a design might be acceptable if its benefits for one person outweigh its risks to that same individual. Risks and benefits to others are not relevant.

This principle is applied in medicine, where a fundamental rule is *non-maleficence*. That is, medical treatments should never be harmful to patients who receive them. Certainly, doctors are not expected to weigh the social consequences of a treatment: It is not a proper concern for a doctor whether or not saving a patient's

---

[13]Bryant (2015).

life would be good for society. Would you want your doctor to consider whether or not the world would be better off without you?

The same principle may be applied to medical devices. Under no circumstances should a medical device be used if it can be expected to leave its recipients worse off.

Note that the same principle does not apply to preventive measures, such as vaccines. A vaccine is not a therapy for a disease, but a means of preventing a disease outbreak from occurring in the first place. As a result, it may be acceptable to mount a vaccine program in a large population, provided that the benefits outweigh the risks overall, even though a few individuals may suffer adverse reactions. In other words, a preventive medical program may be justified on the ground of the common good.

The point is that any medical device intended for therapeutic and not preventive use is expected to observe the principle of individualism.

Case Study: Smart Pacemakers

Pacemakers are implants that help to regulate the operation of people's hearts where they are in danger of heart attack, for example. Many modern pacemakers are able to send and receive information like other wireless devices. In effect, these pacemakers are a part of the "Internet of things". This connectivity is useful because pacemakers can provide information to doctors about the health and functioning of patients' hearts and also allow pacemakers to be reprogrammed without the stress of surgical replacement as circumstances warrant.

However, whenever an item is connected to the Internet, it becomes a potential target for hackers. This point is true of pacemakers. A hacker who gained control of a pacemaker could deliver shocks to a patient's heart that might result in severe injury or death. There has already been an episode of the TV series Homeland in which hackers attack the pacemaker of the US Vice President.[14]

The US Department of Homeland Security has been investigating this risk. Their aim is to help manufacturers identify security risks in their designs and to resolve them. The US Food and Drug Administration has also proposed guidelines for security of Internet-connected medical devices, including pacemakers.[15]

Even so, the advantages offered by these devices are considered to outweigh the risks. Nathaniel Paul, a researcher who has studied these devices, says that the risks are being addressed and that not having the devices at all (or using older models) would be worse.[16] Someone who has a heart attack and is not fitted with a

---

[14]Ryan (2012).
[15]Hsu (2014).
[16]Peck (2011).

pacemaker is likely to die. Older pacemakers help to prevent further heart attacks but are difficult to monitor and can only be repaired or upgraded through major surgery. Pacemakers that can be accessed remotely can be monitored more often and more carefully, and software problems can often be corrected without surgical intervention.

Q: What other risks do medical devices expose patients to? Why are they acceptable, or not?

For medical devices like pacemakers, then, the individualist principle of risk distribution is the correct one to apply.

References

Aven, T. (2012). The risk concept—Historical and recent development trends. *Reliability Engineering & System Safety, 99,* 33–44.

Bryant, R. (2015, January 20). *Jaguar's cars could "tap drivers on the shoulder" to prevent cyclist deaths.* Retrieved January 21, 2015, from Dezeen Magazine: http://www.dezeen.com/2015/01/20/jaguar-bike-sense-alert-tap-drivers-prevent-cycling-accidents/

Bryson, B. (2010). *At home: A short history of private life.* New York: Doubleday.

Hansson, S. O. (2013). *The ethics of risk: Ethical analysis in an uncertain world.* London: Palgrave.

Hsu, J. (2014, October 27). *Feds probe cybersecurity dangers in medical devices.* Retrieved June 20, 2016, from IEEE Spectrum: http://spectrum.ieee.org/tech-talk/biomedical/devices/feds-probe-cybersecurity-dangers-in-medical-devices

Huang, H., Chin, H., & Heng, A. (2006). Effect of red light cameras on accident risk at intersections. *Transportation Research Record: Journal of the Transportation Research Board, 1969,* 18–26.

Keizer, G. (2010). *The unwanted sound of everything we want: A book about noise.* Philadelphia: PublicAffairs Books.

Kusch, L. (2014, September 19). *Diagnosing our health care.* Retrieved October 20, 2015, from Winnipeg Free Press: http://www.winnipegfreepress.com/local/diagnosing-our-health-care-275704401.html

Norman, D. (1990). The "problem" of automation: Inappropriate feedback and interaction, not "over-automation". In D. E. Broadbent, A. Baddeley, & J. T. Reason (Eds.), *Human factors in hazardous situations* (pp. 585–593). Oxford: Oxford University Press.

Norman, D. (2015, June 4). *Automatic cars or distracted drivers: We need automation sooner, not later.* Retrieved June 6, 2015, from Linkedin: https://www.linkedin.com/pulse/automatic-cars-distracted-drivers-we-need-automation-sooner-norman

Peck, M. E. (2011, August 12). *Medical devices are vulnerable to hacks, but risk is low overall.* Retrieved June 20, 2016, from IEEE Spectrum: http://spectrum.ieee.org/biomedical/devices/medical-devices-are-vulnerable-to-hacks-but-risk-is-low-overall

Ryan, M. (2012, December 2). *'Homeland': Brody helps Nazir kill someone; Producers talk shocking exit and what's next.* Retrieved June 20, 2016, from The Huffington Post: http://www.huffingtonpost.com/2012/12/02/homeland-brody-kills_n_2213510.html

Tenner, E. (1997). *Why things bite back: Technology and the revenge of unintended consequences.* New York: Alfred A. Knopf.

The Economist. (2013, November 30). *Fatal distraction.* Retrieved December 3, 2013, from The
  Economist: http://www.economist.com/news/technology-quarterly/21590762-mobile-phones-
  people-who-use-their-phones-while-driving-are-causing-carnage

Whitten, S. (2016, May 25). *Man reportedly caught sleeping behind the wheel of a self-driving
  Tesla.* Retrieved May 27, 2016, from CNBC: http://www.cnbc.com/2016/05/25/man-
  reportedly-caught-sleeping-behind-the-wheel-of-a-self-driving-tesla.html

Fairness

Abstract Moral assessment of designs may proceed by analysis of risk distributions to which they give rise. A complementary approach involves an application of
the concept of *fairness*. Here, fairness refers to moral problems involving the resolution of conflicts of interest between social groups. Such a conflict of interest
occurs when a gain for one social group amounts to a loss for another one. Conflicts
of this type may arise from the operation of designs, especially ones that are widely
adopted. A traffic-routing app that sends drivers through residential streets may
work out well for drivers but less so for residents, for example. In this chapter, the
nature of such fairness problems is described using Taylor-Russell diagrams. These
diagrams help analysts to think clearly about such conflicts of interest. Then, the
fairness impact assessment is defined to describe how Taylor-Russell diagrams may
be employed to develop fair resolutions to such problems.

Introduction

In the previous chapter, we discussed the concept of risk as a means of moral
assessment of designs. The concept of risk is important because it provides a means
of assessing designs when their impacts are uncertain. Designs serve to distribute
risk in a society, so assessment of those distributions allows us to assess designs.
The principles of collectivism, equity, and individualism were described for this
purpose.

In this chapter, we take up a related concept, fairness. Although fairness has
many meanings, the meaning that is relevant here concerns how conflicts of interest
between social groups may be appropriately resolved. Different groups within
society may find that their interests are in competition with the interests of other
groups. In this case, an interest is a stake that social group may gain or lose
depending on how a given situation turns out. A conflict of interest occurs when
two social groups are in a situation where a gain for the interests of one group
means a loss for the interests of the other. Fairness is achieved when competing
interests are balanced appropriately.

C. Shelley, *Design and Society: Social Issues in Technological Design*,
Studies in Applied Philosophy, Epistemology and Rational Ethics 36,
DOI 10.1007/978-3-319-52515-0_11

Designs may give rise to issues of fairness because they often influence how people's interests are satisfied or frustrated. Designs are configured to serve the interests of some social group, usually their clientele. Where those interests are in conflict with the interests of another social group, then a problem of fairness may result. In order to assess designs in such circumstances, any fairness issues that arise should be identified and considered.

In this chapter, we look into how conflicts of interest can arise from designs and how they may be usefully represented. In particular, the Taylor-Russell diagram is developed as a tool for this purpose. After that, the Fairness Impact Assessment, based on the Taylor-Russell diagram is explored as a means of responding to fairness problems implicit in many designs.

Case Study: Light Alert

To see how fairness problems can arise from designs, consider the case of so-called ghetto-avoiders. These are app services that typically combine crime statistics with travel directions. The purpose of these apps is to help users avoid ghettos or bad parts of cities. Examples include apps named SafeRoute, SaferRoute, Road Buddy, and Ghetto Tracker (later re-named to "Good Part of Town"). Indeed, Microsoft has patented "Pedestrian Route Protection" technology that determines walking directions in view of "weather information, crime statistics, demographic information."[1]

It is easy to scorn designs that seem to disguise prejudice as a safety issue. (Imagine an app that helps Engineers to avoid Arts students, or vice versa, on a university campus!) However, some applications are more serious. For a straightforward example, consider LightAlert.[2] This app was designed by a group of female Indiana University students for Microsoft's Imagine Cup competition in 2010. The app tracks the location of the phone and generates an alert if the phone comes within a certain distance of an occurrence of assault or rape, as recorded in publically available police statistics. The alert includes a list of incident reports, the locations of which can be plotted on a map.

The developers support their design by noting some statistics relevant to women on campuses, such as the fact that 20–25% of college women in the United States are victims of assault or attempted assault. A recent study concluded that there were over 700 sexual assaults on Canadian campuses over the last five years, a figure that investigators consider to be an underestimate.[3]

Helping vulnerable people to avoid assault is a serious matter. However, the goal of the design presents some social challenges, even if the design works perfectly from a technical standpoint.

---

[1]Cf. Thatcher (2013).

[2]Schomer (2010).

[3]Ward (2015).

Q: What kinds of mistakes would LightAlert make? What are some consequences of these errors for the people involved?

One sort of mistake would be to warn users in locations where they are not at risk of assault. That is, LightAlert would alert users not to enter an area even though the area is not, in fact, intolerably risky. Conversely, another sort of mistake would be to fail to alert users to areas that are, in fact, intolerably risky.

The first sort of error would tend to stigmatize people who live in an area flagged as dangerous by LightAlert. In effect, it would be considered a ghetto by users. Businesses within the area would loose potential customers that they might otherwise attract. Housing values might also decline as potential residents pass the area over thinking that it is too dangerous.

The second sort of error would tend to endanger users unwittingly. Not until enough users are assaulted would the situation change. It is also conceivable that potential assailants would be attracted to the area if they discover that it is not flagged by the service.

Fairness

For our purposes, fairness means achieving an appropriate balance between the legitimate and competing interests of different social groups. It is clear from the discussion above that LightAlert raises a conflict of interest that calls for appropriate balance. On the one hand, there are the users whose interest is to remain safe. On the other hand, there are the residents of certain areas whose interest lies in how their reputation is affected by public perceptions. The interests of both groups are legitimate, meaning that they are authentic and deserving of consideration. (The interests of potential assailants, who might use the app to target "safe" locations, are harmful and illegal and thus not legitimate. So, they need not be considered in this context.)

This conflict of interest arises from LightAlert's design due to some assumptions that it makes about how to categorize places accurately as safe or unsafe.

Q: What assumption does LightAlert make to perform accurately?

LightAlert considers proximity to a past assault to be a good predictor of present risk of assault. This assumption is plausible as a generality but it is difficult to say exactly how close a person has to be to a past assault to be considered at risk. Similarly, there is an issue of latency. After what interval does a past assault no longer indicate a present danger?

As suggested above, fairness involves balancing such conflicts of interest in a way that respects all legitimate interests of the groups involved. As suggested by the example of LightAlert, fairness problems may arise from mistakes, that is, when assumptions in a design happen to be inaccurate. In order to deal with fairness issues in design in a general way, we will develop a visual tool, the Taylor-Russell diagram, as a way of representing how inaccuracy in the assumptions of designs can lead to conflicts of interest between social groups.

Predictions and Accuracy

As Herbert Simon pointed out, because our knowledge of the world is limited, designs are often based on assumptions. In particular, we do not know the future. As a result, designs often rest on assumptions about the future, which are called predictions. When those predictions turn out to be wrong, a mistake or error has occurred.

The accuracy of such a prediction may be assessed by observing how frequently it turns out to be true and how frequently it turns out to be mistaken.[4] Graphically, accuracy may be plotted in the following way. For each particular prediction that is made, an actual outcome is measured. In a weather forecast, a prediction of 5 mm of rainfall may be checked against the actual outcome, say, 8 mm. These two data can be plotted as the point (5, 8) in a scatterplot, a simple, two-dimensional space. Repeating this process many times produces a scatterplot with many points. Such a plot helps to reveal the accuracy of the predictions.

The accuracy of predictions can be measured quantitatively as a correlation, a number between zero and one. A correlation of one means that the method is predicting the future perfectly, that is, with complete accuracy. A correlation of zero means that the method is predicting the future no better than random guessing.

Graphically, accuracy is represented through the spread of the points in the scatterplot. Consider the smallest possible ellipse drawn around all the points in the plot, with its major axis along a 45° diagonal line. If this ellipse is large, then the points are spread out and accuracy is low. If the ellipse is small, then the points are close together and accuracy is high.

A scatter plot representing a low correlation of 0.2 is given in Fig. 1, a scatterplot representing a medium correlation of 0.5 is given in Fig. 2, and a scatter representing a high correlation of 0.8 is given in Fig. 3.

---

[4]This paradigm is adapted from (Hammond 1996).

Fig. 1 A scatterplot
representing a correlation of
0.2, not very high

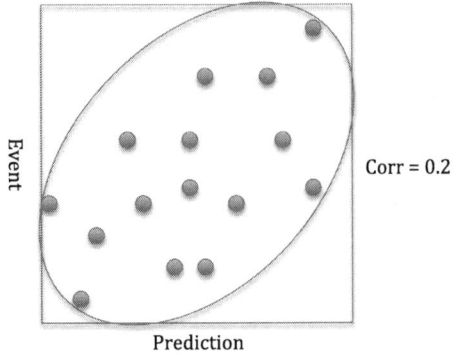

Fig. 2 A scatterplot
representing a correlation of
0.5, a fairly good one

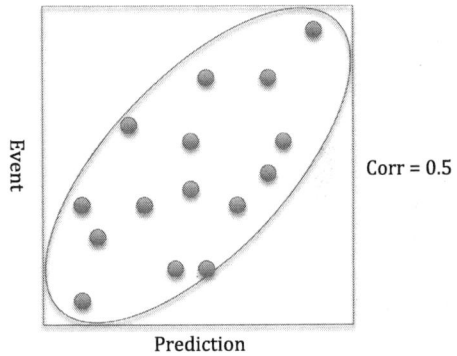

Fig. 3 A scatterplot
representing a correlation of
0.8, a strong one

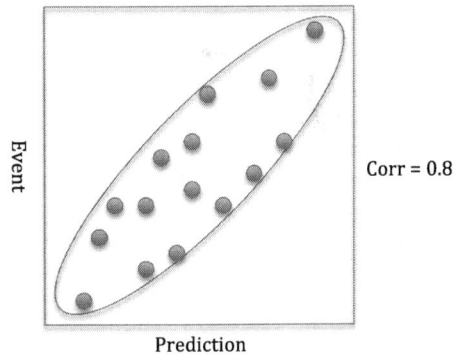

Q: What would the ellipse look like if the correlation is 1?

If the prediction method is perfect then all the points would fall on the line x = y,
since each prediction and observation would be the same.

Fig. 4 A scatterplot of
predictions against events,
featuring a prediction cutoff
(*vertical line*) and a design
threshold (*horizontal line*)

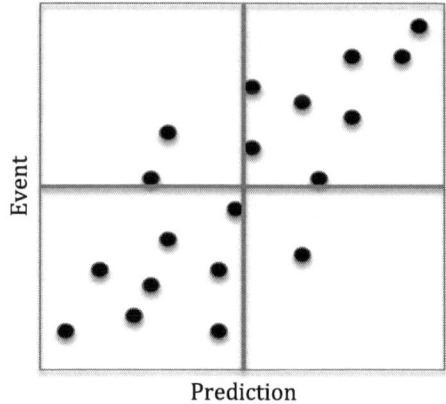

The Taylor-Russell Diagram

Overall accuracy of a prediction system is important but the scatterplot can tell us
even more about the behavior of the system. For this purpose, the scatterplot can be
augmented with two lines to become a Taylor-Russell diagram. Taylor and Russell[5]
were psychologists who studied the use of standardized testing, such as Scholastic
Aptitude Tests (SATs), as a means of determining admissions to university.

Assume that a university prefers to admit students who will proceed to graduate
over students who would not finish their degrees. Since student performance cannot
be known in advance, tests like the SAT are used to predict the likelihood that any
given student will graduate.

The SAT is a test that generates predictions that can be checked against actual
outcomes and plotted as described above. In the scatterplot, the prediction is a
student's SAT score and the event is that student's graduation average. Typically, a
university program requires a test score of at least a certain amount for admission.
Similarly, it requires an average of a least a certain number in order for students to
gain their diplomas. Each of these quantities falls at a point along their respective
axes.

The minimum graduation average is an example of a design threshold. See
Fig. 4. Only students who exceed this threshold may graduate. Graphically, the
threshold may be represented as a line drawn through a scatterplot horizontally
from that point on the event axis. All points above that line represent students who
graduate. All points below it represent students who do not.

Similarly, the minimum SAT score for admission is an example of a prediction
cutoff. Only students who exceed this score may be admitted. Graphically, the
cutoff may be represented as a vertical line through the scatterplot from that point
on the prediction axis. All points to the right of that line represent students whose

---

[5]Taylor and Russell (1939).

Fig. 5 A simple
Taylor-Russell diagram, with
each quadrant individually
labeled

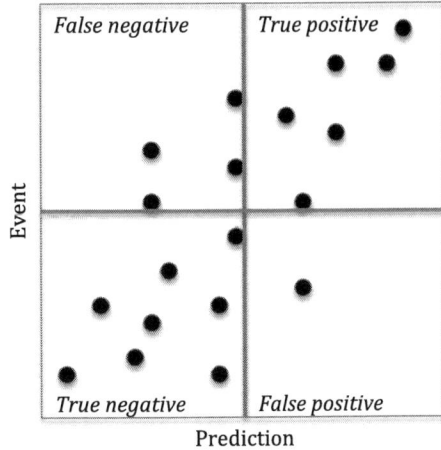

SAT scores allow them to be admitted. All points left of it represent students whose scores do not warrant their admission.

Hammond points out that the situation represented in the Taylor-Russell diagram may be applied to any system in which predictions are used to make decisions about future events.[6] In other words, this method of analysis can be applied to designs in general.

Kinds of Results

The Taylor-Russell diagram can be used to comprehend how predictive successes and mistakes result from designs and what impact those mistakes may have on people. Note that the threshold and cutoff lines in Fig. 4 divide the plot into four quadrants. Each quadrant corresponds to a particular kind of success or failure in the relation between prediction and event, and has a special significance.

First, the upper-right quadrant represents cases where an appropriate decision was taken. Consider the SAT example. Every point within this quadrant represents a student who scored well enough for admission and then went on to graduate from University. This result is known as a _true positive_: Positive because the prediction gave a positive result, e.g., for admission instead of non-admission, and true because the prediction came true. In this case, the system worked as it should. See Fig. 5.

Second, the lower-left quadrant represents cases where the opposite decision was taken. In the SAT example, every point in this quadrant represents a student who

---

[6]Hammond (1996).

did not score well enough for admission and would then not proceed to graduate. (Of course, if students are not admitted, then whether or not they would graduate can only be inferred indirectly, perhaps through comparisons with universities having lower admissions standards. For our purposes, it does not matter how this data is obtained. What matters is only what it means.) This result is known as a *true negative*: Negative because the prediction gave a negative result, e.g., no admission, and true because the prediction was true. In this case also, the system worked as it should.

In the remaining cases, mistakes occur.

Third, the lower-right quadrant represents cases where a decision was taken but did not produce the expected result. In the SAT example, every point in this quadrant represents a student who was admitted to university but did not go on to graduate. This result is known as a *false positive*: Positive because the prediction gave a positive result, e.g., admission, but false because the prediction was false. In this case, the system did not work as desired.

Fourth, the upper-left quadrant represents cases where the opposite decision was taken and did not produce the right result. In the SAT example, every point in this quadrant represents a student who was not admitted to university but who would have graduated if admission had been given. This result is known as a *false negative*: Negative because the prediction gave a negative result, e.g., no admission, but false because the prediction was false. In this case also, the system also did not work as desired.

In any system of less than perfect accuracy, all four kinds of result will occur. The issue is only how they will be distributed.

Tradeoff Between Errors

The Taylor-Russell diagram also reveals an interesting and inextricable link between the two kinds of errors described. Assuming that the accuracy of the prediction system remains fixed, and the design threshold remains unchanged, then the only way to reduce one kind of error is to move the prediction cutoff in the appropriate direction, left or right. However, any move that decreases one kind of error simultaneously increases the other kind.

Consider the SAT example again. Suppose that universities decide that they are admitting too many students who do not go on to graduate. Suppose further that they cannot improve the accuracy of the entrance test. In that case, the remaining strategy to achieve their goal of fewer failures is to move the prediction cutoff to the right, that is, to raise the minimum score needed for a student to be offered admittance to university. With a higher entrance score, fewer academic non-performers are likely to be admitted.

This move makes the lower-right quadrant—false positive—smaller, which means that fewer predicted successes turn out to be failures. However, this move also makes the upper-left quadrant—false negative—larger. As a result, more students will be refused an offer of admission even though they would have graduated successfully.

Here is where a conflict of interest becomes visible. Anyone configuring the system must respond to the question: Which kind of error would it be better to make? The answer, of course, is that it depends upon the value that is placed on each kind of mistake. As Hammond points out, different social groups will value each kind of error differently.[7] As a result, they would advocate for the kind of error they prefer to accept at the expense of the kind that they prefer to avoid.

For instance, some people will argue that a University education (in a public institution) is too expensive for taxpayers to give to many students who will not go on to graduate. This constituency will focus on minimizing false positives and advocate that the cutoff should be moved to the right in order to reduce them. In other words, admission standards should be raised.

However, other people will argue that a University education is important for the economic competitiveness of the country. Raising admission standards means lowering the number of graduates and thus reducing the number of educated workers available nationally in a given field. Such a situation would put the country behind other countries with less stringent policies. This constituency will focus on minimizing false negatives and advocate that the cutoff should be moved to the left in order to reduce them. In other words, admission standards should be relaxed.

In short, a constituency develops focused on each type of error. Each constituency is motivated by a different set of interests that determine their preferences. It is in their interest to see one sort of error decreased. So, they pressure policy makers to change the relevant prediction cutoff in the favored direction.

What people often do not realize is that types of errors are linked in the way discussed above. Thus, a change that promotes the interests of one constituency often comes at the expense of the interests in another constituency, which eventually pushes back. The result is a swing of the cutoff value back and forth over time as each group becomes dissatisfied and motivated by the successful advocacy of the other group.

Error and Fairness

So, a design decision can result in the interests of one group being pitted against the interests of another group. This contest results because errors that occur with the predicted behavior of the design affect people's interests differently. The interests of one social group will be more affected by one kind of error, whereas the interests of a different social group will be more affected by the conflicting kind of error.

---

[7] Hammond (1996).

Politically, the result is a battle for control, to establish a consensus around which sort of error to favor. In terms of social values, the situation gives rise to an issue of fairness. Recall that fairness concerns how the interests of social groups are balanced against one another. It is unfair to favor the interests of one social group if there is nothing about that group that justifies the special treatment. By the same token, it is unfair to disfavor the interests of another social group if there is nothing about the group that justifies this imposition.

Of course, fairness problems of this sort can be mitigated if the accuracy of the relevant predictions can be increased, thus resulting in fewer errors for everyone. Often, however, this is not possible when it comes time to decide on implementing one design or another. So, the question remains: Which distribution of errors is fair? How much should a design serve the interests of one group at the expense of another one?

Fairness and LightAlert

Consider how to apply this concept of fairness to the example of LightAlert discussed earlier. The event that LightAlert is concerned with is the risk of assault. The prediction it uses to model this risk is the distance from a past assault. For the sake of simplicity, we can distinguish between distances that are "small" and would trigger an alert, and distances that are "large" and would not trigger an alert.

This situation can be represented in the T-R diagram in Table 1. Note that the points have omitted so as to focus on the quadrants.

Consider each quadrant in turn. In the upper-right quadrant, the alarm distance is large and the area is high risk. So, the user is kept well away from the dangerous area. This result is a true positive. In the lower-left quadrant, the alarm distance is small and the area is low risk. So, the user is correctly allowed to proceed into a non-dangerous area. This result is a true negative.

In the lower-right quadrant, the alarm distance is large but the area is low risk. In this case, the user is erroneously warned that a safe area is dangerous. This result is a false positive. In the upper-left quadrant, the alarm distance is small but the area is high risk. In this case, the user is not warned against entering a dangerous area. This result is a false negative.

Table 1 T-R diagram describing a fairness issue with the LightAlert app

		Prediction: distance from past assault	
		Small	Large
Event: Risk of assault	High	Danger encountered	Danger avoided
	Low	Safe area labeled	Safe area mislabeled

Fairness Impact Assessment

The Taylor-Russell diagram provides a means for investigating and clarifying problems of fairness that may arise from designs. The diagram allows systematic consideration of errors that designs make and how these may be distributed across social constituencies. The implications of these distributions for social fairness can then be considered.

In order to arrive at a considered judgment of fairness, the Taylor-Russell diagram for a given design needs to be set in context. This context can be obtained by answering a set of questions about the design, the people affected by it, and which principle of fair distribution is most appropriate for the circumstances. This framework of questions can be called a *Fairness Impact Assessment* (FIA).

The framework consists of identifying answers to the following questions:

1. What conflict arises from errors of the design (T-R diagram)?
2. What constituencies are adversely affected by this conflict? How?
3. What social interests are at stake in this case?
4. How could the conflict be resolved fairly?

The first question prompts the development of a T-R diagram to investigate relevant features of a design and whatever errors may arise from its operation. The second question focuses attention on the people who are affected by each kind of error and, more specifically, on what interests they have at stake. The third question invites us to consider the family of problems that are related to the one under consideration. Answering this question draws attention to similar situations that occur with other designs, which may be of use in addressing the fairness issue at hand. The fourth question requires us to consider principled ways in which a design may be configured so as to be fair to the constituencies involved.

Examples of FIAs are given below in order to illustrate this method of analysis.

FIA: LightAlert

LightAlert provides a good place to start.

Consider again the T-R diagram for LightAlert in Table 1. The conflict here results from uncertainty in identifying areas as safe or unsafe for the purpose of providing navigational directions. This conflict may be explicated by characterizing the errors types that LightAlert may make:

- False positive: Some areas that are low risk will be labeled as dangerous.
- False negative: Some areas that are dangerous will be labeled as low-risk.

Low-risk areas mislabeled as dangerous will tend to be considered ghettos. (Thus the nickname "ghetto-avoider" for this genre of service.) Residents' interests will be adversely affected. Business owners may find that customers are scared away from their stores when they might otherwise pay a visit. Residents will find their property values decrease as potential buyers or renters are scared away or demand discounts to take the risk of living there. If the area is an ethnic neighborhood, then mislabeling it may confirm prejudices about its residents among authorities or the general public.

High-risk areas mislabeled as dangerous will tend to expose users of the service to assault, which is exactly what they are seeking, quite reasonably, to avoid.

In general terms, LightAlert raises a security issue. In this case, it pits the safety interests of one group against the economic and reputational interests of another one. Public security systems often raise conflicts of this type.[8]

Answering this question requires us to provide some reason that justifies a certain setting of the design cutoff. In other words, what is the appropriate balance between false positives and false negatives in view of the interests and issues at stake? There are many different ways in which such a justification might be framed, including many principles that we have touched on earlier in this book.

Perhaps the most obvious principle that might be applied is *equality*. That is, configure the design of LightAlert so that it makes an equal number of each kind of error. This approach has the obvious attraction that it puts the same weight on everyone's interests.

This feature can also be a problem. It is not self-evident that everyone's interests should have the same weight. For example, if one group is at a particular disadvantage in this situation, then perhaps its interests should be given special consideration.

Consider this question: How far should providers of the service go to ensure equality? For example, suppose that the service was found to be producing more false positives than false negatives. Would it be acceptable to adjust it so that it does the opposite for a while, and so deliberately exposes more women to assault, in the case of LightAlert, just to make things even? If not, then equality is not the appropriate principle for configuring the service.

Another concept that could be applied is *collectivism*. Recall that this principle states that a design is a good one to the extent that it distributes risk for the best overall outcome. This concept could be operationalized in the LightAlert case by configuring the service so that it produces the fewest complaints from all parties involved. After all, a low level of complaints implies a high level of happiness or satisfaction with the service overall. Plus, risks to all parties are considered.

One problem with this approach is that people who refrain from complaining about a product or service are not always happy with it. For example, residents of an area labeled as a "ghetto" by LightAlert may be simply unaware of what is going on. Unless LightAlert goes to some lengths to advertise their service to such residents, then those people are not really in a position to complain.

Even if residents become aware of the service, they may be less inclined to complain of errors than its users. They may feel that the service providers are unlikely to listen to them, since they are not paying customers. They may feel that

---

[8]Cf. Shelley (2011).

complaining will simply draw more adverse attention to them, thus adding to the problem rather than lessening it.

One principle that is often applied to fairness problems is *proportionality*. That is, the number of errors to which each group is exposed should be proportional to some other relevant quality that applies to both. For example, since women are at a higher risk of assault than the general population, then their interests should be given proportionally more weight. In the case of LightAlert, this approach would mean favoring reduction of false negatives over false positives.

People living in areas in proximity to past assaults might also argue that they are disadvantaged. Such areas are sometimes characterized by economic poverty and low social mobility. The action of apps such as LightAlert is apt to increase this problem.

The principles of equality, collectivism, and proportionality are all aimed at describing a fair distribution of interests directly. A different approach might instead aim to describe not a fair distribution but a fair way of arriving at a distribution. One such principle is *due process*. Due process describes a process for distributing errors (in this case) that is open and unbiased. An open process is one that allows participation by everyone with an interest in its outcome. An unbiased process is one that does not discount or overemphasize the interests of anyone involved in it.

In the case of LightAlert, a due process might consist of a kind of crowdsourced dispute resolution mechanism. Anyone affected by the service might have a say in how it is configured. For example, Google Street View provides broad access to photos of the places where people live and hang out. Their privacy policy requires them to blur out faces and car license plates. However, they will also blur entire cars, houses, or people on request. This policy is popular with celebrities but is available to anyone.[9] Perhaps LightAlert could have a similar process where, say, people around college campuses could weigh in on the risk assessments produced by the service, thus helping to improve the outcome.

Clearly, such an endeavor would be a challenging undertaking for a small group of programmers but perhaps it is required for them to produce a good design.

Q: Which solution is best?

Privacy

Many services available today rely on the collection of personal information about people. A service might collect data to construct a user profile in order to tailor news or advertising to suit that user's tastes and preferences. Since people's interests are often affected by personal information about them, such services often

---

[9]Kleinman (2014).

raise privacy issues. Before examining a particular service, let's take a moment to consider this situation.

There are many treatments of privacy but it is instructive to consider privacy as having at least two components: *confidentiality* and *control*.[10] Having confidentiality means that data about a person is minimally associated with them. In an anonymous database, for example, personal information may be associated with an ID number rather than a name. Then, even if the database is made public, the data remains private in the sense that only the owner of ID number should know that it is associated with them.

Having control means that personal data is disclosed only with permission of the person whom it belongs to. For example, the European Union's "right to be forgotten" states that Europeans have a right to censor online search results for their personal information when that information is "inadequate, irrelevant, or no longer relevant."[11] Google removed nearly one half million links from search results in Europe after requests from users based on this law in its first year. This data is not confidential since it is publically available and results from searches against a person's name. Instead, it is private because the person it applies to can prevent people from finding it.

There are many reasons why privacy may be valuable to people. For one thing, privacy allows people to engage in *impression management*.[12] This term comes from social psychology and refers to how people present themselves to others in order to leave others with a certain impression—usually positive—of them. For example, users of Facebook may spend a great deal of time editing their personal profiles in order to give a positive impression of themselves to their Facebook friends. In this respect, privacy allows people to promote their own interests in the social realm, which is a significant part of getting ahead in life.

By the same token, privacy can facilitate misbehavior. People can use it to hide information about themselves and thus deceive others for their own gain. Consider the Volkswagen diesel emissions scandal of 2015. Software in Volkswagen car engines essentially detected and cheated on emissions tests, to make emissions appear lower than they were under actual driving conditions. One factor that aided the company to cheat for many years was that privacy laws forbade outsiders from examining the software in question.[13]

Since privacy is important to both individuals and society, concerns about privacy often arise as fairness problems.

---

[10]Gurses (2014).

[11]Laurson (2015).

[12]Cf. Tedeschi (1981).

[13]Grimmelmann (2015).

Case Study: Ratsit.se

An informative illustration of issues of privacy versus transparency comes from the introduction of a Swedish Website called Ratsit.se. This website began to publish detailed financial information about Swedish citizens, obtained from the National Tax Board. Sweden has a tradition of openness about the earnings of its citizens, and allows any citizen to check on the basic tax records of any other. So, if you ever wanted to know what your boss makes, or whether your boyfriend is in debt, a trip to the Tax office would fill you in.[14]

Ratsit.se merely obtained these tax records from the Tax office and made them searchable online. So, customers could just input a search at their Web browser, instead of making the trip to the Tax office.

However, the ease with which these records were made available led to a lot of snooping, giving rise to a strong backlash.

"There's a big difference between sitting hidden at home and being reasonably anonymous, and trotting off to the tax office and... telling a person eye-to-eye whom you want to check," said Karolina Lassbo, a 27-year-old lawyer.

Ms. Lassbo said she used Ratsit once "because I wanted to see what it said about me." But her curiosity got the better of her: "Then I checked friends and celebrities."

...

The Data Inspection Board was inundated with complaints, "like an avalanche," said Mr. Karnlof.

The openness that Swedes have towards their tax records was based on important social goals that it could serve. In particular, transparency about incomes can further the goal of income equality. Employees use the information in wage negotiations: An employee might demand a certain wage because tax records show that others with similar jobs make that wage. In that way, Sweden moves towards a condition of equal pay for equal work, an important aspect of economic justice.

However, the radical openness provided by Ratsit.se would pose some dangers. For example, ready access to people's tax records could help identity thieves to steal from them.

For these reasons, the National Tax Board took action. The Board persuaded Ratsit.se to charge for searches on its service. Also, the person whose records are checked is notified of who checked on them. The desired effect is to deter casual snooping, and also to block would-be identity thieves who might use the information to assume a person's identity for fraudulent purposes.

At this point, we are in a good position to complete the FIA.

A T-R diagram representing a fairness issue concerning the Ratsit.se service is given in Table 2. In this case, the event is the nosiness of Swedes, that is, their tendency to casually snoop on the tax information of their fellow citizens.

---

[14]Cf. Nordstrom (2007).

Table 2 T-R diagram of a fairness issue concerning the Ratsit.se service

		Prediction: charge for access to service	
		Low	*High*
Event: Nosiness of Swedes	*Excessive*	Snooping allowed	Snooping deterred
	Moderate	Economic justice facilitated	Economic justice frustrated

Qualitatively, this tendency can be classified as excessive or moderate (acceptable). The prediction involved is the charge made for searches on the service. Charges can be divided by cost into two categories, low and high, with high costs being sufficient to deter casual use.

The conflict here concerns how pricing can be made to allow for legitimate use of the tax information while sufficiently discouraging inappropriate use. This conflict may be explicated by characterizing error types as follows:

1. False positive: High charges will prevent some Swedes from using the information for the purpose of increasing pay equity.
2. False negative: Low charges will allow nosy Swedes to spy on neighbors, acquaintances, and celebrities.

False positives will tend to act against the interests of social groups that tend to experience pay inequity, that is, groups that tend to receive less pay than others for a given type of work. Such groups typically include women and minorities. So, it is these groups that would be disadvantaged most by high costs for searches on Ratsit. se.

False negatives will tend to apply mostly to those must susceptible to spying, which includes politicians and celebrities. Already much in the public eye, these people may feel persecuted if very many fellow citizens choose to look up their tax information.

The Ratsit.se service raises an issue of individual privacy versus a social good, namely economic justice. Both are values that states are bound to defend, so the issue of this trade-off cannot be avoided.

As noted above, the Tax Board hit on an interesting solution. They forced Ratsit. se to raise the cost of searches and to inform people whose records were searched of the identity of the searchers. In effect, the Tax Board had the service reconfigured so that it simulated the previous situation.

In the prior situation, searching records required the searcher to go to the Tax office and fill out some forms. The time and effort needed for this process formed a barrier against casual snooping that Swedes seemed to be happy with.

In addition, the prior situation involved a certain amount of embarrassment. As Ms. Lassbo put it, it is somewhat embarrassing to admit to a Tax official "eye-to-eye" that you wish to inspect someone else's tax records. This embarrassment also formed a barrier against casual snooping. By having the identity of the Ratsit.se user revealed to those whose records they searched, the new situation restores that social cost and thus the barrier it posed.

Since Swedes were apparently happy with the prior situation, then they should also be happy with the new situation that simulates it.

Health

Health care is being profoundly affected by advances in big data and ubiquitous sensing. That is, sensors are becoming cheap enough that they can be embedded almost anywhere and used to collect large amounts of data. Because computers are now so powerful, and so completely networked together, it is possible to access and process mountains of data generated by such sensors. Knowing what do to with the data is not always so easy. The current paradigm favors collecting as much data as possible and figuring out how to use it—that is, "monetize it"—later on.

This paradigm can affect how people view their health. Using smart phones or wearable devices, people can monitor themselves, and transmit and store the data for analysis and comparison. It can be shared with a doctor at a remote hospital. One reason for doing so would be to produce the timely diagnosis of ailments, such as melanoma. Early detection of medical problems would allow for cheaper and more agreeable interventions, and the prevention of serious illness.

The alternative, seemingly more suited to the previous era of less information-intensive medicine, would be the look out for serious conditions and mitigate their effects when they come to threaten health.

Skin Scan

A number of apps have been marketed that use the cameras on smart phones to allow users to monitor any skin growths, tracking them over time and relating them to databases of cancerous growth patterns. One such app is Skin Scan. This app prompts users to take regular photos of any skin abnormalities and monitors their growth over time. The point is to provide a diagnosis of cancer in cases that the app deems suspicious[15]:

> The apps give a recommendation after comparing photos taken of the suspicious lesion over a period of time to gauge unusual changes, or by judging a photo against a library of skin cancer images.

One concern common to services of this type is with their accuracy. A study undertaken by the Cancer Council of Western Australia (CCWA) found that four smartphone apps misclassified more than 30% of dangerous melanomas as safe.

---

[15]Preston (2013).

Table 3 A T-R diagram of a fairness issue concerning skin cancer detection apps

		Prediction: cancer risk score	
		Low	High
Event: cancer risk	Unacceptable	Patient mistakenly assured	Patient correctly alerted
	Acceptable	Benign blemish diagnosed	Benign blemish misdiagnosed

One study of Skin Scan in particular reported the following performance when applied to 93 images of known, dangerous melanomas:[16]

> In our investigation, the sensitivity of Skin Scan to report a melanoma as high risk was 10.8% (10/93). The app classified 88.2% (82/93) of the melanomas as medium-risk lesions and 1.2% (1/93) of the melanomas were reported to be low-risk lesions.

About 11% of melanomas were unclassifiable by the app.

Dr. Terry Slevin of the CCWA argues that this would give users a false sense of security. He agrees that such an app could be useful, particularly for people without regular access to medical care, that is, in remote settings.

At this point, we can complete the FIA.

A T-R diagram representing a fairness issue concerning Skin Scan is given in Table 3.

In this case, the event is the true cancer risk posed by any given skin blemish. Qualitatively, this risk may be divided into acceptable and unacceptable levels. The prediction involved is the risk score calculated by the app for any given skin blemish. For the sake of simplicity, risk scores may divided into high levels (requires immediate attention) and low levels (no immediate attention recommended).

The conflict here concerns how calculated risks scores prompt users to consult their doctors for actual medical conditions without wasting medical resources or endangering patients. This conflict may be explicated by characterizing error types as follows:

1. False positive: Doctors are obliged to correct misdiagnoses of melanomas by apps. Users are also unnecessarily alarmed.
2. False negative: Patients receive inappropriate assurance that their melanomas are benign. Proper medical treatment is then delayed until it is more costly, painful, and perhaps less effective.

False positives will tend to cause unnecessary use of the users' health care system. This use may cause delays for other patients or require more medical resources for the health care system as a whole. Both situations will increase the cost of healthcare for all users, a concern for insurers of the system.

False negatives will tend to primarily affect patients with melanomas. Besides patients themselves, doctors may well find these errors most egregious. This

---

[16]Ferrero et al. (2013).

possibility is reflected in the fact that early studies and commentary on skin care apps by doctors concentrate principally on false negatives.

Apps like Skin Scan enter into a social conflict between the interests of health care managers on the one hand and health care providers on the other. Managers are tasked with running a health care system that delivers the most health care for the amount of money available. Health care providers are tasked with promoting the best interests of each patient.

Dr. Slevin comments that the false negative rate of skin care apps is so high that their use should be restricted to people who have little to no access to a medical care system in any case. Presumably, this recommendation is justified because the false negative rate for such people is already very high, that is, they are very unlikely to be diagnosed with melanoma in any event.

This recommendation appeals to the principle of equity: no one in a society should be given access to medical skin care less effective than can be provided with a simple smart phone app.

However, people who enjoy regular access to a decent health care system should get care at least as good as part of their regular regimen.

References

Ferrero, N. A., Morrell, D. S., & Burkhart, C. N. (2013). Skin scan: A demonstration of the need for FDA regulation of medical apps on iPhone. *Journal of the American Academy of Dermatology, 68*(3), 515–516.

Grimmelmann, J. (2015, September 22). *Harry Potter and the mysterious defeat device.* Retrieved September 28, 2015, from Slate.com: http://www.slate.com/articles/technology/future_tense/2015/09/volkswagen_s_cheating_emissions_software_and_the_threat_of_black_boxes.html

Gurses, S. (2014). Can you engineer privacy? *Communications of the ACM, 57*(8), 20–23.

Hammond, K. R. (1996). *Human judgment and social policy: irreducible uncertainty, inevitable error, unavoidable injustice.* New York: Oxford University Press.

Kleinman, A. (2014, July 7). *How to remove your house from Google Street View.* Retrieved June 27, 2016, from Huffington Post: http://www.huffingtonpost.com/2014/07/07/remove-google-street-view_n_5563939.html

Laurson, L. (2015, April 23). *How Google handled a year of "right to be forgotten" requests.* Retrieved April 24, 2015, from IEEE Spectrum: http://spectrum.ieee.org/telecom/internet/how-google-handled-a-year-of-right-to-be-forgotten-requests

Nordstrom, L. (2007, June 17). *Swedes revolt against online snooping.* Retrieved June 22, 2007, from The Washington Post: http://www.washingtonpost.com/wp-dyn/content/article/2007/06/17/AR2007061700468.html

Preston, R. (2013, January 13). *Skin cancer apps 'dangerous'.* Retrieved January 20, 2013, from The Sydney Morning Herald: http://www.smh.com.au/digital-life/smartphone-apps/skin-cancer-apps-dangerous-20130117-2cva6.html

Schomer, S. (2010, October 24). *Student app "LightAlert" warns users of rape zones.* Retrieved October 26, 2010, from FastCompany: http://www.fastcompany.com/1597217/student-app-light-alert-warns-users-rape-zones

Shelley, C. (2011). Fairness and regulation of violence in technological design. *International Journal of Technoethics, 2*(4), 20–36.

Taylor, H. C., & Russell, J. T. (1939). The relationship of validity coefficients to the practical effectiveness of tests in selection: discussion and tables. *Journal of Applied Psychology, 23*(5), 565–578.

Tedeschi, J. (Ed.). (1981). *Impression management: Theory and social psychological research.* London: Academic Press.

Thatcher, J. (2013). Avoiding the Ghetto through hope and fear: an analysis of immanent technology using ideal types. *GeoJournal, 78*(6), 967–980.

Ward, L. (2015, November 23). *Schools reporting zero sexual assaults on campus not reflecting reality, critics, students say.* Retrieved November 24, 2015, from CBC News: http://www.cbc. ca/news/canada/campus-sexual-assault-survey-1.3328234

Progress

Abstract One of the central features of good design relates to the concept of progress. That is, people expect that designs will improve over time, so that later designs tend to be better than earlier ones. Technical progress is often associated with innovation, as discussed earlier. However, there is also the matter of moral progress, the expectation that the world will become more ideal over time. Contributions that designs may make to moral progress may be instructively considered by adapting the concept of fairness from the previous chapter. On this view, progress may be seen as a moral dilemma between two strategies for regulating designs, that is, *permissive* and *precautionary* strategies. The permissive strategy recommends that new designs be accepted for general use unless and until they prove to be harmful. The precautionary strategy recommends that new designs be restricted from general use unless and until they prove to be safe. Relationships between, and institutional attitudes towards, these two strategies are discussed.

Introduction

In the previous chapter, we discussed issues of fairness. A problem of fairness arises when two social groups have a conflict of interest and serving the interests of one group means detracting from the interests of the other. As we have seen, designs can involve problems of fairness because of how they serve the interests of one group or another in some way. Resolving a fairness problem often involves finding an appropriate balance between the kinds of errors that a design is likely to make.

In this chapter, we will adapt this fairness concept to a related issue in design assessment, namely progress. For present purposes, progress has to do with how designs change over time. Normally, we would say that progress has been made if designs improve as time goes on. For example, we expect that newer versions of a line of computers or smartphones will be better than the older ones, perhaps working at a faster speed or with more memory.

Besides this notion of technical progress, there is also the notion of *moral progress*. Moral progress implies that the world becomes an increasingly ideal place

© Springer International Publishing AG 2017   191
C. Shelley, *Design and Society: Social Issues in Technological Design*,
Studies in Applied Philosophy, Epistemology and Rational Ethics 36,
DOI 10.1007/978-3-319-52515-0_12

in which to live. Perhaps society will increasingly approach some ideal form in which people are able to thrive and get along with each other as well as can be. Herbert Simon describes the matter this way[1]:

> Moral progress has always been associated with the capacity to respond to universal values
> —to grant equal weight to the needs and claims of all mankind, present and future.

This statement also reminds us of Dieter Rams's view, discussed earlier on, that the social mission of design is to make the world a more humane place. How can we judge whether or not a design is helping to accomplish this goal?

Determining whether or not a new design conforms to a social ideal or makes the world more humane is quite challenging. A simpler but still worthwhile determination of the same type can be made by adapting the concept of fairness developed earlier. We can consider the simplified question: Would it be fair to accept or to reject a new design or design feature? Or, should the innovation be accepted but only under certain restrictions? These questions are considerably complicated by the fact that any effects of a new design are often largely uncertain. As a result, reasonable people may arrive at very different answers.

In this chapter, we address the problem of moral progress in design by adapting the fairness concept developed in the preceding chapter. This adaptation will allow us to understand an important dilemma faced by regulators who assess technology under conditions of significant uncertainty. The dilemma concerns the best strategy to following in order to give the public access to the benefits of innovation while protecting them from undue harm.

Case Study: Brain Stimulation

One innovative category of design is consumer brain stimulators. In brief, these stimulators are devices that are fitted on people's heads and affect their moods or states of mind by changing their brains.

An important category of brain stimulator works on a principle known as Transcranial Direct Current Stimulation (TDCS). TDCS works through induction of currents in the brain through electrodes placed on a person's scalp. Since brain operation depends on electrical signals between neurons, the state of a person's brain can be changed by manipulating those signals with induced electric currents. In basic terms, an induced current can help to prime or intensify activity in a brain structure or it can help to suppress it. By positioning electrodes appropriately, any brain structure might be targeted, allowing users to affect any sort of brain process.[2]

Medical research using TDCS has shown that it can reduce pain, ease depression, treat autism and Parkinson's disease, control cravings for alcohol and drugs,

---

[1]Simon (1981), p. 184.
[2]The Economist (2015).

repair stroke damage, and accelerate recovery from brain injuries, to say nothing of improving memory, reasoning and fluency. Remarkably, some effects seem to persist for days or even months.

Do-it-yourself hackers have been inspired by this research to use similar equipment to experiment on their own brains[3]:

> Christopher Zobrist, a 36-year-old entrepreneur based in Vietnam, is one of them. With little vision he has been registered as blind since birth due to an hereditary condition of his optic nerve that has no established medical treatment. Mr Zobrist read a study of a different kind of transcranial stimulation (using alternating current) that had helped some glaucoma patients in Germany recover part of their vision. Despite neither the condition nor the treatment matching his own situation, Mr Zobrist decided to try TDCS in combination with a visual training app on his tablet computer. He quickly noticed improvements in his distance vision and perception of contrast. "After six months, I can see oncoming traffic two to three times farther away than before, which is very helpful when crossing busy streets," he says.

Enthusiasts have set up websites to exchange stories and design ideas.

Brain researchers have expressed reservations about DIY brain hacking. Peter Reiner of the National Core for Neuroethics at the University of British Columbia says that incorrect placement of electrodes or direction of current might impair the processes that users are trying to boost.[4] In addition, there is little to no data suggesting how TDCS interacts with other brain stimulants like coffee or marijuana or what effects it might have on people with conditions such as epilepsy. Some neuroscientists are concerned about potential effects of TDCS on the neural development of children and young adults.[5]

Thync

Perhaps the best-known commercial brain stimulator is Thync. Thync is a small device typically worn over a user's right temple. It generates a patterned electrical field that affects cranial nerves and brain structures in that area.

The device is controlled by a smartphone app and operates in two modes: *calm* and *energizing*. One reviewer described the effect of calm mode in this way[6]:

> The calm mode (or "calm vibes," as Thync describes it) left us feeling us a bit like we'd just smoked a joint, and the energy mode led to more of a stimulated clarity—as if a mental fog we weren't even aware of had been lifted.

The same reviewer tried the energizing mode, which apparently produced a state of behavioral vigor:

---

[3]The Economist (2015).
[4]The Economist (2015).
[5]Cf. Wurzman et al. (2016).
[6]Shanklin (2015).

Fig. 1 Aaron Muszalski
wearing Thync/Wikimedia
commons. URL: https://
commons.wikimedia.org/
wiki/File:I,_Lobot._—_
Taking_the_@Thync_
#neurosignaling_wearable_
for_a_spin_at_@runway_is._
(2015-07-02_19.08.37_by_
Aaron_Muszalski).jpg

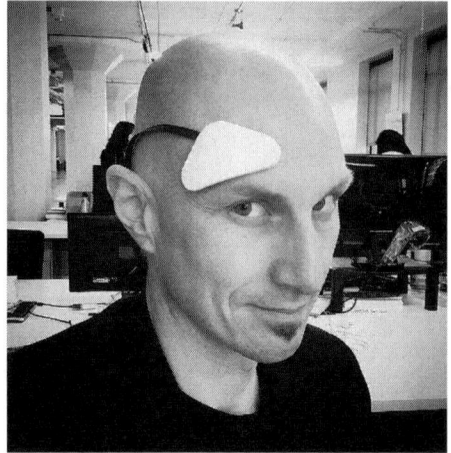

After a short break, I used Thync again in the energizing mode. These changes weren't immediately obvious, but they became evident in my behavior. After energizing, I was talking more often – and more loudly – with greater expression and animation. This mode felt a bit less like I was slipping into a different state of mind, but it affected me nonetheless.

Each mode can be adjusted according to the intensity of the mood desired.

Other than similar reviewer impressions, very little is known about any effects of Thync. A small study performed by Thync's designers suggested that the device's calm mode may decrease physiological arousal a bit more than a placebo treatment.[7] No other studies of the device have been published in the scientific literature.

However, Thync was introduced not as a medical device but as a recreational one. As such, it needs to meet only relatively relaxed safety standards. Since it is a new kind of device, it is not clear what it would mean for Thync to be considered safe. Obvious possibilities would be that it does not burn its users or produce thoughts of suicide or murder.

So, a significant problem for regulators is that there is very little information about what effects Thync may have on people, especially in the long term. To see the significance of this problem, it helps to put yourself in the shoes of regulators who might be considering how to react to the invention of Thync (Fig. 1).

Q: How Thync be regulated? Restrictively, loosely, or in-between?

Dr. Peter Reiner of the National Core for Neuroethics at the University of British Columbia argues that regulation of such devices should be light.[8] In his view, since

---

[7]Tyler et al. (2015).
[8]The Economist (2015).

such devices are cheap and can be constructed by DIY hackers for as little as $20, heavy regulation would simply increase prices of commercial products and encourage the use of unregulated or underground designs.

The Concept of Moral Progress

It is clear that Thync represents a technical innovation. The issue that remains is what level of access to Thync would be best. This situation illustrates how assessment of *moral* progress is separate and additional to an assessment of *technical* progress.

The distinction between technical and moral progress may be clarified as follows. In general terms, progress occurs when things get better. In other words, progress means that things in the future are better than they were in the past. Here, the term *better* means something like "more good". That is, progress means that things in the future rate more highly in our assessment of good than they did in the past.

Note that this concept of progress is ambiguous. Recall from our earlier discussion that *good* in *good design* can have two meanings:

1. Rational: Good designs are ones that achieve their ends in excellent ways, whatever those are.
2. Moral: Good designs are ones that help to achieve excellent ends.

Since we have defined progress in terms of good, it follows that the concept of progress is ambiguous in the same way. Consider the following characterizations of progress in design in terms of designs becoming "more good":

1. Rational: Progress is made when designs become more excellent in the achievement of their ends, whatever those are.
2. Moral: Progress is made when designs help to achieve more excellent ends.

Thync is arguably an example of rational progress. It may well be better than drugs or meditation, for example, in achieving easy and effective control over moods. Even if this claim is true, it may not be the case that such control is an excellent ability to acquire, at least under all circumstances. Thus, it remains unclear whether or not Thync is an example of moral progress.

A Dilemma of Progress

As noted above, one way to assess moral progress in an innovation is to consider whether it would be fair to accept or reject the innovation or, more realistically, to regulate it loosely or restrictively. This issue is significant for innovations like

Thync where something substantial is at stake, such as people's brains, and where there is substantial uncertainty about how such innovations will affect people.

The problem for regulators can be considered in the following way. Regulators face the difficulty of constructing a response to innovations that have the potential to produce significant risks for the public. Risks are often distributed to the public unevenly, such that the interests of some constituencies are advanced while others are set back. Also bearing in mind how uncertain the impacts of a new design will be, regulators may settle on relatively stringent or relatively loose restrictions on its introduction to the marketplace.

The nature of this fairness problem facing regulators can be clarified by performing a fairness impact assessment, beginning with construction of a Taylor-Russell diagram in Table 1. This diagram is the same as those discussed in the fairness chapter, except that the prediction of interest for regulators concerns the level of regulation that should be applied in a given case.

In this construction, the event of interest is the level of safety risk posed by Thync to the public. Relevant risks might include damage to their heads or unhealthy alterations in their moods. Viewed qualitatively, these risks may be either acceptable or excessive (and thus unacceptable).

The prediction of interest for regulators, as mentioned above, is the level of regulation that should be applied. Viewed qualitatively, regulation may be restrictive or loose. Restrictive regulation of Thync might be to consider it as a medical device, in which case a great deal of study would be necessary in order for regulators to approve of it as safe for general use. Loose regulation would consider only protection of consumers from burns or other obvious physical damage.

The false positives in the lower-right quadrant relate to safety concerns posed by Thync users to the general public, which may be similar to those posed by drug or alcohol use. The false negatives in the upper-left quadrant relate to the loss of potential benefits of Thync for users, which may include greater happiness and increased cognitive ability.

The social issue at stake is a familiar one. It pits the interests of individual users who may thrive better with access to devices like Thync than they would otherwise against the rights of the public to live free of problems posed by people whose brains or moods have been damaged by brain stimulators.

Then there is the issue of resolving the matter fairly. One commentator argues for minimization of false negatives on grounds of proportionality, because devices

Table 1 A T-R diagram representing a progress problem posed by tDCS devices like Thync

		Prediction: regulation of Thync	
		Restrictive	Loose
Event: safety risks	Acceptable	Users denied mood control *false negative*	Users enjoy mood control
	Excessive	Public spared safety risks	Public exposed to safety risks

false positive

such as Thync will help a social group that is in particular and substantial need of them[9]:

> Consumer-wellness devices like Thync may appeal to those who cannot use caffeine or alcohol for medical or religious reasons, and there will always be healthy overachievers seeking to supercharge their cognition for study or work. More importantly, TDCS presents the tantalising promise of relief from some medical conditions for which traditional therapies are either ineffective or unaffordable. As the University of Melbourne's Mr. Horvath says, "If there are ten percent of people who are feeling a huge effect, even if that's placebo, who are we to say no to them?"

This argument leaves out concerns about impacts on potentially vulnerable populations such as children and young adults. If these concerns are valid, then perhaps Thync could be regulated along the lines of alcoholic drinks, with age restrictions applied.

Permissive and Precautionary Strategies

difficult choice has to be made between 2 or more ↑ alternatives

As the Thync case illustrates, the problem for progress poses a dilemma. In general, this dilemma takes the form of two competing strategies that may be applied to deciding on how restrictive access to the new design should be. On the one hand, access could be made quite broad. As in the old saying "seize the day", the idea would be to regulate a new design loosely and thus realize any benefits that it brings as quickly as possible. On the other hand, regulations could be more restrictive. As in the old saying "it is better to be safe than sorry", the idea would be to thoroughly study a new design before marketing it and thus avoid any harms that it might bring about.

These two strategies fall into general categories:

1. The *permissive strategy* enjoins *(urge)* us to press ahead with the adoption of new technologies, to seize the day. By following this strategy, we may obtain the benefits of all innovations that we develop.
2. The *precautionary strategy* enjoins us to restrict adoption of technologies unless we can be quite sure that they will not cause us major harms. By following this strategy, we will spare ourselves the harms of inappropriate innovations, along with the costs of cleaning up the damage they do.

The two strategies are at odds in the sense that both cannot be applied at once: If one of them is followed, then the other one is not. In addition, each strategy has a roughly equally persuasive rationale to support it. Since there is typically substantial uncertainty surrounding the consequences of adopting or not adopting a new design, advocates of either strategy can appeal to the same set of facts in order to support their recommendation. This dilemma is a genuinely difficult one.

---

[9]The Economist (2015).

Table 2 A generalized T-R diagram representing the relationship between permissive and precautionary strategies for moral progress

		Prediction: accept design	
		No (precaution)	*Yes* (permissive)
Event: value realized	*Positive*	Forgo benefit	Gain benefit
	Negative	Avoid harm	Incur harm

This dilemma can be captured in a general way by the following T-R diagram (Table 2).

The precautionary strategy is focused in false positives, that is, instances in which we adopt a technology that turns out to inflict harm. That error is the one to avoid on this strategy. The permissive strategy is focused on the false negatives, that is, instances in which we decide not to adopt an innovation that would, in the end, have provided a worthwhile benefit. Advocates of permissiveness urge that this kind of error is the one most to be avoided.

Supporters.

Case Study: VapShot

Inhalation has become a novel way to self-administer intoxicants. Into this arena comes a new product named VapShot that allows users to inhale their drinks. VapShot is a machine that gasifies an alcoholic beverage and deposits the vapor in a plastic bottle so that users can inhale it through a straw. When the bottle is uncorked, the gas pops like a bottle of champagne!

According to its designers, VapShot has several advantages.[10] First, alcohol is absorbed through the blood vessels of the lungs, thus taking effect right away.

Since the alcohol is not imbibed, it does not pass through the stomach and intestines where it sometimes makes people feel sick to their stomachs. In addition, vaporizing beverages also helps to bring out their flavor, the makers claim.

On the whole, the company claims that VapShot provides a more efficient and pleasurable way to get drunk.

The machine comes in two forms. The first is the *VapShot mini* that is meant for consumer use and costs $700. There is also a commercial version meant for bars that can serve 720 shots per hour and sells for $4000.

Of course, any device designed to intoxicate people raises safety issues.[11] Victor Wong, the CEO of the company, argues that VapShot has resolved these issues. He argues that VapShot has been designed to moderate alcohol intake by limiting the amount of alcohol in each shot. According to tests commissioned by the company,

---

[10]VapShot, n.d.

[11]Nguyen (2014).

subjects who used the machine to take in a dose of hard liquor (80 proof) every 10 min for an hour had, within their lungs, levels of ethyl alcohol that were well below where a regulatory agency like the Occupational Safety and Health Administration (OSHA) would consider unsafe for use in workplace.

Also, he points to animal studies done in the 1970s suggesting that inhaled alcohol is eliminated from the body more quickly than imbibed alcohol. Again, according to company data, Breathalyzer readings that saw blood alcohol levels in subjects who had one VapShot drop from a peak of 0.05% to undetectable over the course of 10 min.

Critics remain concerned about VapShot. One concern is that users may not be able regulate their alcohol intake effectively. When people drink too much, their digestive system responds with an urge to vomit, thus removing excess alcohol that is poisoning them. Since VapShot avoids the digestive tract in favour of the lungs, this mode of auto-regulation is bypassed. So, VapShot users may not realize how drunk they are and may even suffer severe alcohol poisoning.[12]

"It is ill advised for experimentation among those under 21,' said Dr. Thomas Greenfield, Center Director at the National Alcohol Research Center in Emeryville, California.

'There could be inexperienced people at parties under peer pressure who may find themselves using this method of alcohol consumption.

'It might not be possible to self-regulate their consumption and teenagers just like adults can be drunk drivers too."

William C. Kerr, a senior scientist at the Public Health Institute's Alcohol Research Group, a nonprofit, remains skeptical.[13] Technologically imposing limits on how much gets inhaled at once, he contends, would ultimately do little to discourage abuse. "Even if you reduce the concentrations, who's to say a user wouldn't use it on top of drinking to enhance the effects," he says. "It's hard to say how people would behave give the chance, but you need to take into account what's more common instead of just what's possible."

In addition, Kerr argues that VapShot may lead to lung damage. He argues that since there has been virtually no research on the effect of vaporized alcohol on people's lungs, it would be better to restrict products like VapShot until the matter is cleared up through further research.

A similar device to VapShot, the AWOL (Alcohol Without Liquor) has already been banned in 22 states for this reason.[14]

---

[12]Prigg (2014).

[13]Nguyen (2014).

[14]Associated Press (2006).

VapShot Dilemma

The policy dilemma posed by VapShot can be clarified following the model outlined above. The first step is to construct a Taylor-Russell diagram to identify what interests are at stake and for whom.

Since we are dealing with a policy problem, the prediction involved is about how restrictive regulation of VapShot should be. In qualitative terms, the prediction cutoff falls between regulations that are restrictive and those that are relatively loose. The event involved concerns various health risks that VapShot may pose. In total, those risks may turn out to be acceptable or excessive (Table 3).

The false positives in the lower-right quadrant relate to concerns that users of VapShot will be given to drunkenness and irresponsible behaviors that follow, such as drunk driving, and medical problems such as alcohol poisoning. In turn, these matters have obvious implications for public safety.

The false negatives in the upper-left quadrant concern the denial of the potential advantages of VapShot to uses in the event that the design poses no unusual risks to them. Perhaps the device does simply provide people with an advantageous way of becoming pleasantly intoxicated.

Waiting until further research is performed would represent a precautionary approach.

Like the case of Thync, the social issue relevant to VapShot concerns individual access to recreational narcotics versus public safety in the face of individuals who use the device unwisely.

> Q: How should VapShot be regulated?

As in the case of Thync, do-it-yourself versions of VapShot are not difficult to construct. So, there is a similar argument for allowing consumer versions to be mass-manufactured with limited regulation. Also as with Thync, there might be some attempt to restrict VapShot to users over a certain age so as to curb abuse by inexperienced persons.

Table 3 A T-R diagram representing a progress problem for VapShot

		Prediction: regulation of VapShot	
		Restrictive	Loose
Event: health risks	Acceptable	Users denied safe alcohol hit *false negative*	Users enjoy safe alcohol hit
	Excessive	Public spared risks of drunkenness	Public exposed to risks of drunkenness *false positive*

Institutional Bias

In considering progress, we have not yet considered the fact that regulators themselves may be biased towards one strategy over the other one. The result is that different authorities may look at very similar situations and come to different decisions about which strategy to follow. Differing tolerances or preferences between regulators is known as *institutional bias*.[15]

Consider a comparison between electrical power utilities in North America versus India.[16] Among the main tasks of a power utility is matching the supply of electricity to demand. Demand may vary depending upon the weather and other external events. Supply is a matter of generating capacity and the ability to apply it when and where it is demanded.

North American utilities tend to keep about 10% excess supply available at any given time. When conditions warrant, such as the occurrence of a heat wave when demand for electricity to run air conditioners may spike, extra generators are kept running so that their power is available to be supplied to the grid in order to satisfy sudden demand increases.

In India, by contrast, the supply of electricity is well below the level needed to meet spikes in demand. This occurs for several reasons. Politically, it is difficult to justify to people in a developing country that a lot of money should be spent on generating capacity that will not be used most of the time. Other, pressing uses exist for that funding. The price of electricity is also controlled and subsidized, so that there is less incentive for power utilities to invest in grid infrastructure. The result is that managers of India's power grids use rolling blackouts to manage situations in which demand outstrips supply. Sometimes, the result is large-scale blackouts from which recovery is difficult. For example, 620 million people in India experienced a black out between July 30 and 31, 2012.

The bias of each electricity provider can be represented by contrasting Taylor-Russell diagrams. In these tables, the status quo, and thus the precautionary strategy, is assumed to be the provision of extra generating capacity beyond expected demand spikes. The event of concern is the level of demand, which is either high or low relative to generating capacity.

The two diagrams are identical in terms of their contents. The difference is in the prediction cutoff in each case. In the American case, the cutoff is pushed left, representing the decision to reduce risk of blackouts being needed to manage electricity supply (the false negative); see Table 4. In the Indian case, the cutoff is pushed right, representing the decision to reduce risk of what could be construed as excessive spending on generating capacity; see Table 5.

Comparison of these diagrams allows us to clarify the dilemma facing power utilities in both countries and also a significant difference in their responses. These

[15]Little (2005).

[16]LaMonica (2012).

Table 4 A T-R diagram representing the North American bias concerning electricity supply

North America		Prediction: provision of extra capacity	
		No	*Yes*
Event: demand	*High*	Blackouts required	Blackouts avoided
	Low	Money saved for other uses	Money kept from other uses

Table 5 A T-R diagram representing the Indian bias concerning electricity supply

India		Prediction: provision of extra capacity	
		Negative	*Positive*
Event: demand	*High*	Blackouts required	Blackouts avoided
	Low	Money saved for other uses	Money kept from other uses

responses are the result of different political and economic priorities in these two jurisdictions.

Case Study: Self-driving Cars

United States officials announced an investigation into the death of Joshua Brown, who was killed on May 7, 2016, as a result of a collision between his Tesla Model S car and an 18-wheeler truck on a Florida highway.[17] The death was notable because it was the first confirmed fatality involving a car running Tesla's Autopilot mode. According to the company, the crash happened because the car's system could not distinguish between the while color of the truck's trailer and the background sky, which prevented the car from braking.[18] Reports suggest that the driver may have been watching a movie and thus did not perceive the danger.[19]

The Autopilot feature of the Tesla Model S is not a fully automated driving system. Instead, it combines adaptive cruise control, automatic steering, automatic lane changes, and automatic emergency steering.[20] So, it can handle most of the

---

[17]Hull et al. (2016).

[18]Tesla Motors (2016).

[19]Reuters (2016).

[20]Oremus (2016).

Table 6 A T-R diagram representing Tesla's permissive bias in driving automation

Tesla		Prediction: accept partial automation	
		No	Yes
Event: safety risk	Acceptable	Driving remains less productive	Driving made more productive
	Excessive	Inattentive drivers saved	Inattentive drivers harmed

Table 7 A T-R diagram representing Google's precautionary bias in driving automation

Google		Prediction: accept partial automation	
		No	Yes
Event: safety risk	Acceptable	Driving remains less productive	Driving made more productive
	Excessive	Inattentive drivers saved	Inattentive drivers harmed

routine tasks of driving in good conditions but still requires drivers to take control under certain, unusual conditions. The result illustrates the partial automation problem discussed under the heading of risk: Since monitoring an automated system is tedious and unproductive, drivers will tend to focus attention on other tasks, leaving them unprepared to take over control in difficult circumstances. This may well explain the circumstances of Joshua Brown's death.

The case also highlights differences of strategy in the development of self-driving cars by different companies. While Tesla has opted to introduce and combine automation features one-by-one, Google has opted to develop an entirely self-driving car first. During tests on its partially-automated models, Google engineers noticed that people's attention would wander and they were unable to control the car properly when the need arose.[21] As a result, Google decided to develop a fully automated car, one without a steering wheel or gas and brake pedals, before entering the marketplace.

We may assume that Google's final product will be relatively safer than the Tesla Model S with Autopilot. (And, both may be safer than unassisted and distracted human drivers.) As such, Google's strategy is precautionary: It involves holding off on acceptance of an innovation until it is proven adequately safe. By

---

[21]Markoff (2016).

contrast, Tesla's strategy is more permissive: It involves marketing an innovation on the assumption that it will prove to be adequately safe in use.

These strategies can be represented and contrasted using T-R diagrams. See Tables 6 and 7.

In reading these diagrams, differences between the strategies become more apparent. Tesla has placed relatively more weight on the productivity of its customers, who may use the attention freed by the Autopilot to read, watch movies, or sleep. In addition, Tesla seeks to grab early marketshare and learn from the volumes of data it obtains from the cars running the Autopilot feature in order to refine it. Perhaps this is why Tesla describes the release of Autopilot as a "public beta phase."

Google has placed relatively more weight on the safety of its customers, who would be under no obligation to monitor the driving behavior of their vehicle. However, because of the delay required to develop such a truly self-driving car, Google denies the public any benefits of whatever partial automation it could introduce earlier, as well as an early chance to gain marketshare and real-world experience with its systems.

Q: Which strategy is better?
Q: What other designs are there that illustrate contrasting biases between permissive and precautionary strategies?

References

Associated Press. (2006, Oct 8). *Citing safety, states ban alcohol inhalers.* Retrieved July 5, 2016, from The New York Times: http://www.nytimes.com/2006/10/08/us/08whiskey.html.

Hull, D., Plungis, J., & Welch, D. (2016, July 1). *U.S. regulators investigating Tesla autopilot death.* Retrieved July 4, 2016, from The Globe and Mail: http://www.theglobeandmail.com/report-on-business/international-business/us-business/tesla-shares-slip-after-first-fatality-involving-autopilot/article30725609/.

LaMonica, M. (2012, Aug 12). *Outage in India could be a harbinger for the rest of the world.* Retrieved Oct 23, 2012, from Technology review: https://www.technologyreview.com/s/428685/outage-in-india-could-be-a-harbinger-for-the-rest-of-the-world/.

Little, R.G. (2005). Organizational culture and the performance of critical Infrastructure: Modeling and simulation in socio-technological systems. In *Proceedings of the 38th Annual Hawaii International Conference on System Sciences* (p. 63.2). IEEE.

Markoff, J. (2016, July 4). *Tesla and Google take different roads to self-driving car.* Retrieved July 7, 2016, from The New York Times: http://www.nytimes.com/2016/07/05/business/tesla-and-google-take-different-roads-to-self-driving-car.html.

Nguyen, T.C. (2014, Aug 15). *Need a buzz? Inventor claims his device offers a safer, gentler way to inhale liquor.* Retrieved Nov 3, 2014, from The Washington Post: https://www.washingtonpost.com/news/innovations/wp/2014/08/15/need-a-buzz-inventor-claims-his-device-offers-a-safer-gentler-way-to-inhale-liquor/.

Oremus, W. (2016, July 6). *Is Autopilot a bad idea?* Retrieved July 7, 2016, from Slate: http://www.slate.com/articles/technology/future_tense/2016/07/is_tesla_s_style_of_autopilot_a_bad_idea_volvo_google_and_others_think_so.html.

Prigg, M. (2014, July 14). *The controversial $700 gadget that lets you inhale alcohol at home - despite health warnings.* Retrieved July 18, 2014, from The Daily Mail: http://www.dailymail.co.uk/sciencetech/article-2695220/The-controversial-700-gadget-lets-INHALE-alcohol-home.html.

Reuters. (2016, July 1). *DVD player in Tesla raises questions in Autopilot death.* Retrieved July 4, 2016, from Huffington Post: http://www.huffingtonpost.com/entry/tesla-dvd-death_us_57770959e4b09b4c43c0912d.

Shanklin, W. (2015, June 2). *Thync mood-changing wearable officially launches—we go hands-on again.* Retrieved June 5, 2015, from Gizmag: http://www.gizmag.com/thync-hands-on-2/37820/.

Simon, H. (1981). *Sciences of the artificial.* Cambridge, MA: MIT Press.

Tesla Motors. (2016, June 31). *A tragic loss.* Retrieved July 4, 2015, from Tesla blog: https://www.teslamotors.com/en_CA/blog/tragic-loss.

The Economist. (2015, March 7). *Hacking your brain.* Retrieved March 13, 20015, from The Economist: http://www.economist.com/news/technology-quarterly/21645509-diy-bundle-electronics-or-ready-made-device-it-possible-stimulate.

Tyler, W. J., Boasso, A. M., Mortimore, H. M., Silva, R. S., Charlesworth, J. D., Marlin, M. A., et al. (2015). Transdermal neuromodulation of noradrenergic activity suppresses psychophysiological and biochemical stress responses in humans. *Scientific Reports, 5,* 1–17.

VapShot. (n.d.). *What is VapShot?* Retrieved July 18, 2014, from VapShot.com: http://www.vapshot.com/what-is-vapshot-.html.

Wurzman, R., Hamilton, R. H., Pascual-Leone, A., & Fox, M. D. (2016). An open letter Concerning do-it-yourself users of transcranial direct current stimulation. *Annals of Neurology, 80*(1), 1–4.

Sustainability

Abstract Any current discussion of good design includes *sustainability*. In its most basic sense, sustainability refers to our ability to maintain and develop our lifestyle or civilization in the long term. In this sense, sustainability may be understood instructively as a progress problem of the type discussed in the previous chapter. That is, sustainability concerns a dilemma over permissive and precautionary strategies in our consumption of resources. The permissive strategy usually focuses on increasing efficiency in individual designs on the assumption that this measure will decrease resource consumption overall. However, Jevons' Paradox tends to undermine this assumption. The precautionary strategy often focuses on *biosynergism*, that is, designs that emphasize both internal integrity and environmentalism. A challenge for this strategy is that it may require changes in consumer lifestyle that people generally will find difficult to accept.

Introduction

Sustainability is an important part of good design. However, it is not self-evident what sustainability means in terms of design. One reason for this lack of clarity is that sustainability poses a moral dilemma of progress of the type discussed in the previous chapter.

For present purposes, *sustainability* is a problem of managing consumption in the face of limited resources. Modern living has involved ever more consumption of resources. Yet, those resources are not boundless. Simply carrying on a consumption-intensive way of living until a crucial resource runs out could lead to economic and social collapse. Sustainability means finding some way of preventing severe social disruptions while still enjoying the benefits that modern living has to offer.

In terms of design assessment, sustainability presents a problem of moral progress. This is because consumption may be regulated in at least one of two ways. The first strategy is to leverage economic growth in order to encourage innovations that would allow for consumption to remain within resource limitations. For example, as innovation makes designs more efficient, those designs consume fewer

© Springer International Publishing AG 2017

C. Shelley, *Design and Society: Social Issues in Technological Design*,
Studies in Applied Philosophy, Epistemology and Rational Ethics 36,
DOI 10.1007/978-3-319-52515-0_13

resources, thus delaying the time when they run out. This approach may be called the *growth strategy*.

The second strategy is to limit consumption itself and thus conserve resources. Conserving resources would postpone the time when they run out. For designers, the challenge is to find ways to limit consumption while maintaining prosperity and acceptable standards of living. This approach may be called the *degrowth strategy*.

The purpose of this chapter is to examine this dilemma and some challenges that it presents for designers. It is a truism today that good design involves sustainability. However, what it means for a design to be sustainable, and how sustainability in general may be enhanced through design, remains difficult to specify with certainty.

Case Study: Shower Heads

A morning shower is a daily ritual for many people in developed countries. In the United States, the Environmental Protection Agency estimates that showering accounts for 17% of household water used, adding up to 1.2 trillion gallons (4.5 trillion liters) annually.[1] Fresh water is a limited resource and in short supply in some areas, so its sustainability is an important matter.

To increase sustainability of water consumption, the US government adopted the U.S. Energy Policy Act of 1992. One feature of this Act was to mandate greater efficiency in shower heads. Up to that time, American shower heads allowed flow rates of 5–8 gallons (19–30 L) per minute. The Act capped flow at 2.5 gallons (9.5 L) per minute.

The idea was straightforward. With shower heads that restrict water flow, showers would become more efficient with water, that is, a shower of a given length could be accomplished using much less water than before. More efficient showers would lower water consumption, thus making the overall water supply more sustainable.

Q: What factors might undermine this plan?

At first, shower manufacturers simply inserted restrictors that lowered water flow into their existing shower head designs.[2] Many consumers found the experience underwhelming, complaining that their showers no longer got them very wet or left them very relaxed.

Also, many Americans felt that the low flow rate did not allow them to wash their hair properly. On an episode of the contemporary TV show *Seinfeld*, after low-flow

[1]Environmental Protection Agency (n.d.).
[2]Ball (2009).

Fig. 1 A low flow shower
head. Photo by John
Loo/Flickr.com. https://www.
flickr.com/photos/johnloo/
3004182709/

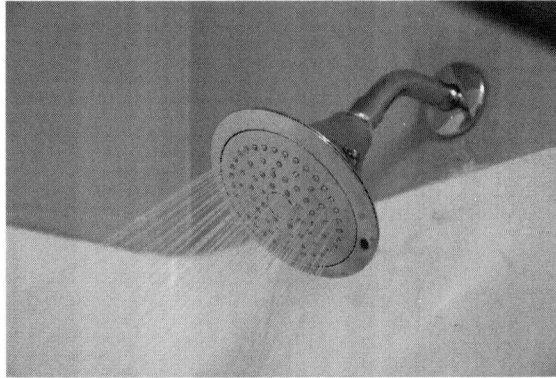

showerheads were installed in their apartments, the characters found that they could
not rinse shampoo from their hair effectively.[3] After a few days of exasperation, they
bought black-market shower heads (the "Commando 450," used in the circus for
elephants) and removed the low-flow designs. In real life, many Americans simply
took the flow restrictors out of their new shower heads. Some manufacturers actually
included instructions showing how the restrictors could be removed.

Americans who stick with the low-flow shower heads sometimes find that extra
shower time is needed to accomplish hair washing or relaxation properly.
Additional time in the shower then reduces efficiencies realized through flow
restriction.

In an effort to finesse the problem through technology, shower head designers
have invented ways to insert air into water droplets made by their goods. Inserted
air increases the size of droplets and simulates the feeling of higher water flow. One
difficulty with this approach is that mixing water with ambient air lowers the water
temperature. In order to preserve the desired shower experience, aerated showers
require water at higher temperatures, which requires about 10% more energy per
shower. This unintended consequence means that this increase in water efficiency
decreases efficiency of energy usage.

Low-flow shower heads may have saved a fair amount of water. However, their
introduction also illustrates some of the problems that can arise in designing
technology for increased sustainability (Fig. 1).

Case Study: Power Drills

In the wake of World War II, American home builders began a program of constructing lots of inexpensive housing for returning soldiers. To be affordable, these
houses lacked finished basements and attics. It became normal for homeowners to

---

[3]Mehlman et al. (1996).

improve their houses during their occupation, such as finishing basements and
attics, by themselves. This practice stimulated the do-it-yourself (DIY) movement,
which remains popular today.[4]

To facilitate the advent of DIY home improvement, tool manufacturers designed
consumer versions of their professional tools, often lighter and less robust versions
of the latter designs. Extensive advertising promoted the view that every real man is
a handyman, with his own set of power tools. As a result, around half of American
households today have power drills and similar tools, which spend the vast majority
of their time on shelves doing nothing at all. It is estimated the average power drill
is used for between six and thirteen minutes in total over its lifetime (Fig. 2).[5]

Collaborative Consumption – *degrowth strategy*

An obvious difficulty with this state of affairs is that having millions of power drills
lying around doing nothing most of the time seems a waste of resources. In place of
this consumerist mode of distribution, Rachel Botsman and Roo Rogers propose
what they call *collaborative consumption*.[6] This idea refers to the sharing of goods
or services within a group of people.

For example, consider the service Neighborgoods.net.[7] It is a service in which
people in a given neighborhood can sign in to list goods that they would like to
borrow from neighbors or that they would be willing to lend to neighbors. The
goods are typically limited to small, portable things, such as bicycles, lawn mowers,
and also power tools.

The system has enjoyed some early successes in U.S. cities such as Los Angeles
and San Francisco. For example, tech-savvy urban farmers with the Phoenix

---

[4]Gelber (1997).
[5]Botsman and Rogers (2010).
[6]Botsman and Rogers (2010).
[7]Ferenstein (2011).

Permaculture Guild share over $2500 worth of wheelbarrows, shovels, and other city-bound agricultural equipment that helps members to access infrequently used items.

Q: What are some other examples of collaborative consumption? What are some trade-offs of this approach?

Public libraries are another example. In the case of public libraries, the pool of books and other media are owned and organized by a city rather than being owned and organized in a peer-to-peer fashion by the members of the service themselves.

Sustainability and Progress

The case of shower heads illustrates a growth strategy for sustainability. In order to increase sustainability, it identifies a particular form of consumption and seeks to redesign it for greater efficiency. The idea is that sustainability in general can be increased by taking any one means of consumption and making it more efficient. One appealing feature of this strategy is that it does not ask people to change established patterns of consumption.

The case of Neighborgoods.net illustrates a degrowth strategy.[8] In order to increase sustainability, it identifies a particular form of excess capacity and seeks to reduce it by re-organizing its use and distribution. Reducing the amount of stuff that people need to accomplish their goals is intended to help conserve resources. One of the more challenging features of this strategy is that it asks people to alter their established patterns of consumption.

Together, these examples illustrate a dilemma of progress in design for sustainability. This general dilemma can be captured in a Taylor-Russell diagram; see Table 1. In this case, the event of interest is risk of a resource crisis, which would include events such as severe droughts or energy shortages, that could undermine the integrity of a society. The prediction of interest is the emphasis on economic growth within that society. In this case, the growth strategy is considered permissive and the degrowth strategy precautionary, as the practice of growth in the past is already generally admitted to be unsustainable for the future.

Proponents of degrowth prefer to minimize false positives, that is, cases where consumption leads to resource crises that threaten to undermine social integrity. In the worst case, a severe drought might lead to a social collapse, as has happened into some societies in the past.[9]

---

[8]Cf. Jackson (2009) and Assadourian (2012).
[9]Diamond (2005).

Table 1 A T-R diagram representing a general progress dilemma regarding sustainability

		Prediction: encourage growth	
		No (precaution)	*Yes* (permissive)
Event: risk of resource crisis	*Acceptable*	Consumers deprived of more goods *deny false* –	Consumers enjoy more goods *True* +
	Excessive	Social integrity maintained *True* –	Social integrity disrupted *False* +

In more modest cases, resource crises might exacerbate social divisions. In Canada, for example, the National Energy Program was a response to the oil supply crises of the 1970s. It gave the federal government power to redistribute oil from the western province of Alberta to the rest of the country at below-market prices.[10] The program created resentment in Alberta, especially towards Canadians in the eastern half of the country, an issue that divides Canadians up to the present day.

Proponents of (mitigated) growth prefer to minimize false positives, that is, cases where consumers are denied benefits of consumption. Having to borrow a power drill from a neighbor rather than fetch one from one's own basement might discourage homeowners from improving their houses, for example. In that case, they would forfeit increases in the value of those houses that may follow from home improvement. More generally, not being able to market new goods broadly in the marketplace might discourage people from innovating new and better designs.

As is often the case with progress problems, the two strategies emphasize different general social interests. The degrowth strategy tends to emphasize the integrity of the social group as a way of increasing sustainability while the growth strategy tends to emphasize provision for the actions of individuals. As these interests are both legitimate, achieving an appropriate balance is a challenging task.

Consumption and Efficiency

The growth strategy tends to equate sustainability with overall efficiency of consumption, and overall efficiency with the sum of the efficiencies of individual designs. To understand this rationale better, it is instructive to examine the relationship of efficiency to sustainability. This examination also allows us to understand some limitations of this strategy.

All resources are limited in supply, that is, no resource is infinite in extent. Thus, there is a need to limit consumption of them in some way. Those limitations can be thought of as restrictions on the rate of consumption. This perspective is embodied in axioms 3 and 4 of Richard Heinberg's five axioms of sustainability[11]:

---

[10]Bregha (2006).

[11]Heinberg (2007), pp. 88–95.

Fig. 3 A representation of
the rate of consumption of a
renewable resource over time

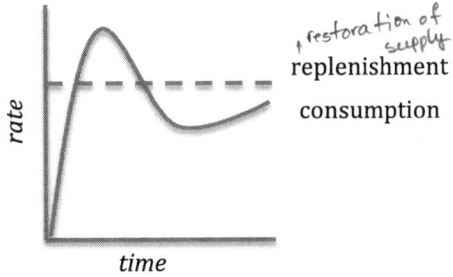

3. To be sustainable, the use of renewable resources must proceed at a rate that is less than or equal to the rate of natural replenishment.
4. To be sustainable, the use of non-renewable resources must proceed at a rate that is declining, and the rate of decline must be greater than or equal to the rate of depletion.

The third axiom could be depicted graphically as in Fig. 3.

If a renewable resource, e.g., lumber, is replenished at a given rate, then consumption is sustainable if it depletes the resource at a lesser rate. A small period of consumption at a higher rate is possible if a reserve of the resource is available; that is, if the area under the consumption curve but over the replenishment line does not exceed the resource reserve.

The fourth axiom could be depicted graphically as in Fig. 4.

To be sustainable, the rate of consumption of a non-renewable resource, e.g., oil, must reach zero before the resource is exhausted; that is, the area under the curve cannot exceed the size of the resource reserve. In effect, this curve is the same as the previous one but with a replenishment rate of zero.

In this framework, sustainability means limiting consumption of resources to rates that do not lead to their exhaustion. Increasing efficiency of consumption helps in achievement of this goal because it defers exhaustion until later. In terms of the figures above, the point of exhaustion moves further right.

Fig. 4 A representation of
the rate of consumption of a
nonrenewable resource over
time

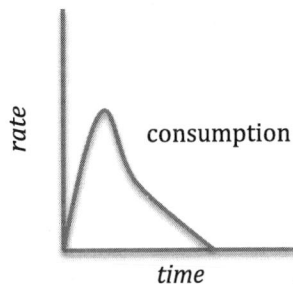

This approach seems compelling because it is focused strictly on the design of technology rather than habits of consumption. Put another way, it promises to achieve sustainability without the need for explicit social activism.

Even so, this focus on efficiency of consumption faces some important difficulties.

Jevons' Paradox

Increasing efficiency of individual designs does not necessarily lead to overall decreases in consumption. This fact was first observed and explained by William Jevons (1835–1882), a British economist, in a book named "The coal question" (1865) (Fig. 5).[12]

Jevons pointed out a seeming paradox of progress in technology. As coal engines became more efficient in design, coal consumption as a whole was actually increasing instead of decreasing. The result was that, because of these increased efficiencies, Britain was going to run out of coal sooner rather than later.

The case of coal itself is instructive. As coal locomotive engines became more efficient, the prices of train tickets fell, putting them within the means of more people. Thus, more train trips resulted. Also, increasing profitability of the industry meant that more railways could be built, further increasing railway travel.

Jevons calculated that Britain would run out of coal by the mid 20th Century, a prediction that turned out to be roughly correct. Coal mining in Britain became infeasible in the 1960s and was shut down in the 1970s.[13]

Since Jevons' time, his paradox has been divided into two issues[14]:

1. The *rebound effect*, in which some, but not all, of the gains in efficiency in some design are eroded by additional consumption; and
2. The *backfire effect*, in which gains in efficiency are totally eroded by additional consumption. This effect was the one that concerned Jevons.

For our purposes, differences between these effects are not too important. However, the backfire effect is clearly the worst news we could have when using efficiency to pursue sustainability goals.

---

[12]Jevons (1865/1965).
[13]Hallett and Wright (2011), pp. 43–54.
[14]Mokhtarian (2009).

Fig. 5 Willam Stanley
Jevons (1835–1882), the
British economist who posited
that increases in efficiency
lead to increases in
consumption. From Popular
Science Monthy,
1877/Wikimedia commons.
URL: https://en.wikipedia.
org/wiki/William_Stanley_
Jevons#/media/File:PSM_
V11_D660_William_
Stanley_Jevons.jpg

Explanations

There is a tendency to assume consumption patterns are largely fixed, so that increasing efficiency of consumption merely allows those patterns to persist for a longer time. That is the implication of Heinberg's axioms. What Jevons discovered is that increasing efficiency of consumption intensifies consumption patterns themselves.

Since Jevons' time, scholars have identified three sorts of ways that Jevons' paradox can occur.[15]

1. *Direct rebound*: Gains in efficiency lower the unit cost of a resource, thus stimulating greater consumption of it.
2. *Indirect rebound*: Gains in efficiency in one resource produce cost savings that end up being consumed through consumption of other resources.
3. *Dynamic rebound*: Gains in efficiency in one resource stimulate innovations in new modes of consumption, resulting in an overall increase in resource demands.

---

[15]Mokhtarian (2009).

A good example of <u>direct rebound</u> would be <u>gains in efficiency in car engines</u>. As engines become more efficient, <u>consumption might be expected to drop. However, people who purchase more efficient cars tend to take more trips</u> or <u>longer ones</u>. Studies estimate that about 1/5 of the fuel savings from increased efficiency in cars is consumed in this way. Even so, direct rebound leaves room for considerable reduction of consumption from gains in efficiency.[16]

In effect, increasing the efficiency of fuel consumption is economically equivalent to discovering more fuel, thus lowering its unit price and making consumption more affordable, thus stimulating demand somewhat.

> Q: What indirect and dynamic effects could result from increased automotive efficiency?

For <u>indirect rebound</u>, as <u>people make more road trips, then more roads will be built or improved. Road construction is energy and resource intensive</u>. Additional equipment and infrastructure is needed as well, such as tires, gas stations, fuel refineries, as well as metals and plastics to build more cars.

As for <u>dynamic rebound</u>, the <u>mobility granted by cars and enhanced roadways lead to new sorts of consumption. Simple examples include drive-thru restaurants, motels, and summer cottages.</u>

It is difficult to calculate the amount of indirect and dynamic rebound caused by efficiency gains in design. How much of cottage construction is due to increases in car engine efficiency? There is no simple way to answer that kind of question. The problem of accounting for increases in energy consumption in one area due to efficiency gains in another is difficult. This difficulty only adds to the uncertainty that comes with the problem of making progress in sustainability.

Jevons' Paradox suggests that the relationship between sustainability and efficiency is more complex that is often thought. A strategy founded on simply making designs more efficient is not guaranteed to work.[17]

Case Study: Lighting

Two of the most basic uses of energy are the generation of light and heat. There is no doubt that generation of light and heat have become much more efficient, especially in recent history.

In the case of light, recent developments in compact fluorescent bulbs (CFL) and solid-state (LED) lighting have made lighting much more efficient. Historically,

---

[16]Hallett and Wright (2011).

[17]Cf. Alcott (2005).

increased efficiency in lighting has been accompanied by more demand for it.[18] Light perceived by the human eye is measured in units called lumen-hours. This is about the amount produced by burning a candle for an hour. In terms of labor, an ancient Babylonian would have to work about 41 h to acquire enough lamp oil to produce 1000 lumen-hours of light. A Briton living around 1800 would have to work only 5 ¼ h to be able to produce the same amount of light, from burning tallow candles. Today, the same person would have to work for only ½ s to produce that light with a CFL bulb. These figures suggest just how much light production has gained in efficiency over time.

In terms of actual light production, a typical Briton living in 1800 produced about 580 lumen-hours of light per year. Today, the same person produces about 46 megalumen-hours, nearly 100,000 times as much.

According to a model developed at Sandia National Laboratories by Jeff Tsao, the introduction of LED lighting will make light generation three times more efficient than with CFLs and yet lead to a ten-fold increase in light generation 2030.[19] On the assumption that LED lighting will be about three times more efficient than CFLs by 2030, and that the price of electricity remains the same in real terms, then Tsao's model predicts that average consumption will jump to 202 megalumen-hours per person. Furthermore, the amount of electricity needed to supply the additional light will more than double. This increase can be prevented only if electricity prices triple in the meantime.

Greater efficiency in electricity generation makes the last scenario seem unlikely.

> Q: What indirect and dynamic effects could result from increased efficiency in lighting?

For indirect effects, cheaper lighting could prompt more interest in things and activities that can be illuminated, such as buildings, signs, vehicles, roadways, and so on. As for dynamic effect, increased lighting could stimulate activities such as travel at night, sports at night, and recreational lighting displays, not to mention modes of consumption such as eating and drinking that might accompany them.

Interestingly, laser lighting may overtake LED lighting with the next decade, as laser lighting is still more efficient. Some of the first uses of laser lighting imagined by promoters include laser light shows at home and at work, and the illumination of airports and entire building exteriors.[20]

It is worth noting that not all increases in efficiency need result in increases in consumption. For example, an Indian Non-Governmental Organization called Mokshda Paryavaran Evam Van Suraksha Samiti invented a more efficient way to cremate bodies for traditional Hindu funeral pyres. The main component of the

---

[18]Nordhaus (1996).

[19]Tsao (2010).

[20]Mims (2013).

design includes a hood that concentrates heat, thus allowing cremation to finish using less wood fuel than before.[21] The increased efficiency of this process may lead to more funerals if those become cheaper but it is unlikely to prompt Hindus to die at a faster rate.

Biosynergism

A growth strategy centered on designing things to make consumption more efficient will not necessary stave off a resource crisis. A degrowth strategy would instead center on re-organizing consumption in order to conserve resources. The appeal of such a strategy would be that it directly addresses the problem of overconsumption. An important problem is that it could require significant changes in lifestyle, which people are apt to dislike or resist.

One approach to implementing degrowth in design of technology may be called *biosynergism*. Biosyngerism refers to a form of bio-inspired design, on which sustainability is enhanced through emulation of natural systems. The concept of biosynergism goes back to Scottish biologist Patrick Geddes (1854–1932) (Fig. 6). Besides being a biologist, Geddes was what we might call an urban planner. He studied the structure of cities and tried to determine what factors distinguished well-organized cities from poorly-organized ones. He used his knowledge of ecology do this, comparing "healthy" cities to healthy ecosystems.[22]

Geddes bio-inspired approach to "healthy" and sustainable cities relies on two central ideas:

1. Organicism: maintaining internal integrity;
2. Environmentalism: maintaining external integrity;

These ideas are discussed below.

Organicism

Organicism is an ecological concept that refers to the so-called web of life. Ecologists observed that populations of organisms exist in a complex network of relationships and mutual dependencies; see Fig. 7. For example, small animals depend upon plants for food. Large animals depend on the small ones in the same way. The manure generated by these animals helps to spread seeds and fertilize the soil, which helps the plants to survive. This arrangement is sometimes what is known as a "food web."

---

[21]The Economist (2007).
[22]Shelley (2016).

Fig. 6 Patrick Geddes
(1854–1932), the Scottish
biologist and urban planner
who promoted a bio-inspired
view of sustainability. URL:
https://commons.wikimedia.
org/wiki/Category:Patrick_
Geddes#/media/File:Patrick_
Geddes_(1886).jpg

Fig. 6 Patrick Geddes (1854–1932), the Scottish biologist and urban planner who promoted a bio-inspired view of sustainability. URL: https://commons.wikimedia.org/wiki/Category:Patrick_Geddes#/media/File:Patrick_Geddes_(1886).jpg

Patrick Geddes coined the term *synergy* to describe how an integrated web of such populations tends to be mutually supporting. The dependency of each node in the network on the other nodes tends to lock all the populations together and helps to ensure the survival of each. For this reason, synergy tends to enhance the sustainability of all the populations because of their support for one another.

In design, synergy means that all components of a design are well integrated with one another. That is, each component must facilitate the operation of others, and no component should hinder the operation of others.

For example, consider the Visitor Centre of the VanDusen Botanical Garden in Vancouver. Designed by Perkins + Will in 2011, the Centre is a place where visitors can experience the amenities of the garden and participate in special events. The architects collaborated with an ecologist consultant in its design.[23] Not surprisingly, it features a number of synergistic design elements[24]:

1. Waste from its toilets and food waste composted from its restaurants are combined and used as fertilizer for the plants in its gardens.

---

[23]Busby et al. (2011).
[24]Flint (2015).

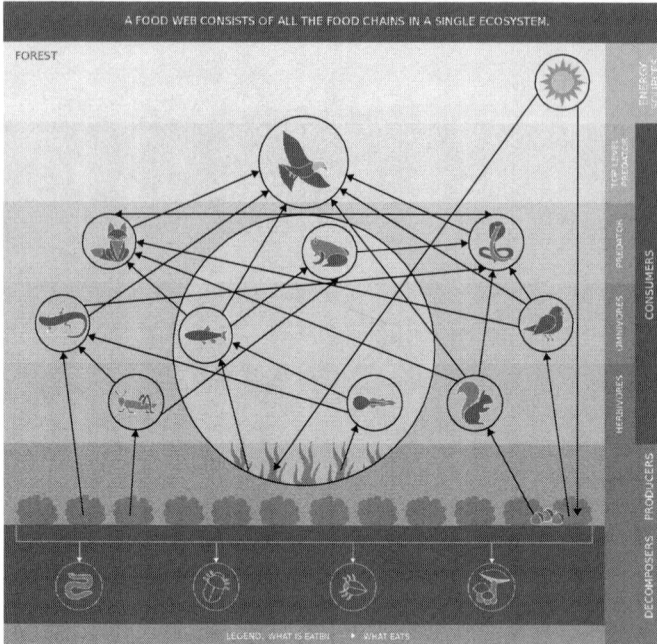

Fig. 7 A food web. Picture by Roberta Rosina, Density Design Research Lab/Wikimedia commons. URL: https://commons.wikimedia.org/wiki/File:Food_Web.svg

2. Wastewater is separated out and purified and used for irrigation in the gardens.

In both of these cases, things that are produced by one system in the Visitor Centre are used as inputs that contribute to the operation of another system. In that way, each of these component helps to sustain other ones. This kind of internal integration results in systems that are as self-sufficient as possible in their operations.

Environmentalism

Organicism concerns how well integrated the components of a system are with one another. Much the same concern can be raised regarding how well a system is integrated into other systems in its surroundings. In fact, this form of integrity represents the aspect of biosynergism known as *environmentalism*. In this sense, environmentalism refers to how well a design operates with other things that it interacts with. A design is well integrated in this sense if it facilitates the functioning of other things that it interacts with and does not hinder them.

For example, a group of MIT engineering students won the first MIT Water Innovation Prize, 2015, with a device called AquaFresco. This device aims to radically reduce the amount of water and detergent used by washing machines.[25]

> Although washing machines are more efficient than ever, they still use more than 20 gallons of water to remove 1 tablespoon of dirt. And in addition to the water use itself is the problem of detergent.
>
> "[Washing machines] are one of the major sources of detergent pollution in rivers," [Sasha] Huang says. "Current laundry technology is not sustainable. A regular washer discards the water right into the drain after one usage, but less than 1 percent is the actual waste component."

AquaFresco filters the wash water to remove dirt and recover fresh water and unused detergent for future use. It is being tested in hotel laundry systems, where water usage is especially intensive.

The environmentalist aspect of AquaFresco is expressed in the way that it minimizes the external resources, that is, water and detergent, that it requires from the outside world. Also, the design minimizes any pollution it generates that would otherwise be discharged into the environment, which it would tend to degrade. By decreasing its input requirements and also any pollution that it outputs, AquaFresco helps to prevent damage to the external environment. These features help washing machines and their environment to be as mutually-sufficient as possible in their operations.

Linear Versus Circular

linear

circular

The difference between consumerist and biosynergistic design has been characterized as a difference between "linear" versus "circular" design by architect John Tillman Lyle.[26] Lyle captured this difference in the form of two diagrams.

linear The first diagram suggests how consumer designs relate to their environment in a linear fashion; see Fig. 8. They take in resources from the external world, use some of them up, and shed waste products, some of which may be harmful to the surroundings. The flow of energy and materials is linear; here, from left to right.

circular The second diagram suggests how biosynergistic designs, which Lyle called "regenerative," relate to their environment in a circular fashion; see Fig. 9. They attempt to replenish external resources that they use, recycle energy and materials internally as much as possible, and release into the environment things that the environment can readily neutralize or put to work without disruption.

This diagram well illustrates the importance of integrity, in the forms of organicism and environmentalism, to the biosynergistic approach to sustainable design.

---

[25]Wolfe (2015).
[26]Lyle (1994).

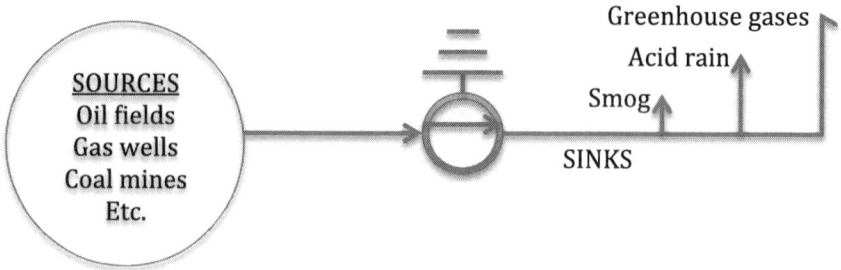

Fig. 8 A schematic representation of a linear design for resource consumption

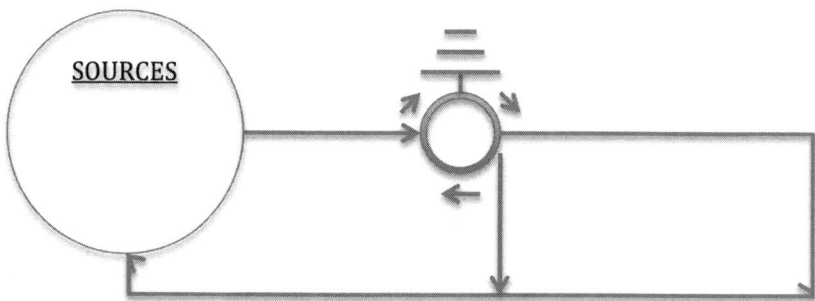

Fig. 9 A schematic representation of a circular design for resource consumption

Q: What other biosynergistic ("circular") designs can you think of? What makes them biosynergistic?

Sustainability is an important aspect of good design. Its main challenge is how to ensure thriving and prosperity in the face of limitations of important resources. In this sense, it can be understood as a problem of moral progress, a choice between strategies of growth and degrowth.

The growth strategy emphasizes thriving through consumption. It deals with limitations on resources by striving to make consumption more efficient, usually in a piecemeal fashion, so that consumption may be sufficiently prolonged. Problems with this strategy include Jevons' Paradox and environmental disruptions from release pollutants.

The degrowth strategy emphasizes thriving through conservation. It deals with limitations on resources by rationing consumption of them, usually through a "circular" approach to design, so that resources remain for future use if necessary. A problem with this strategy is its requirement for potentially profound changes in people's established lifestyles.

In terms of sustainability, a design may be assessed by identifying which of these strategies it fits with and whether or not that strategy is the appropriate way to consume limited resources while solving a problem.

References

Alcott, B. (2005). Jevons' paradox. *Ecological economics, 54*(1), 9–21.

Assadourian, E. (2012). The path to degrowth in overdeveloped countries. *Moving toward sustainable prosperity* (pp. 22–37). Washington: Island Press.

Ball, J. (2009, November 13). *Under pressure: Bathers duck weak shower heads.* Retrieved July 11, 2016, from Wall Street Journal: http://www.wsj.com/articles/SB125807041772846273

Botsman, R., & Rogers, R. (2010). *What's mine is yours: The rise of collaborative consumption.* New York: Harper Business.

Bregha, F. (2006, July 2). *National Energy Program.* Retrieved July 11, 2016, from The Canadian Encyclopedia: http://www.thecanadianencyclopedia.ca/en/article/national-energy-program/

Busby, P., Richter, M., & Driedger, M. (2011). Towards a new relationship with nature: Research and regenerative design in architecture. *Architectural Design, 81*(6), 92–99.

Diamond, J. (2005). *Collapse: How societies choose to fail or succeed.* New York: Viking Press.

Environmental Protection Agency. (n.d.). *Showerheads.* Retrieved July 11, 2016, from United States Environmental Protection Agency: https://www3.epa.gov/watersense/products/showerheads.html

Ferenstein, G. (2011, June 2). *NeighborGoods aims to be a prettier, more social, community-generating Craigslist.* Retrieved June 5, 2011, from FastCompany: http://www.fastcompany.com/1742262/neighborgoods-aims-be-prettier-more-social-community-generating-craigslist

Flint, A. (2015, May 19). *Can regenerative design save the planet?* Retrieved May 23, 2015, from The Atlantic Citylab: http://www.citylab.com/design/2015/05/can-regenerative-design-save-the-planet/393626/

Gelber, S. M. (1997). Do-it-yourself: Repairing and maintaining domestic masculinity. *American Quarterly, 49*(1), 66–104.

Hallett, S., & Wright, J. (2011). *Life without oil: Why we must shift to a new energy future.* Amherst, N.Y.: Prometheus Books.

Heinberg, R. (2007). *Peak everything: Waking up to the century of decline in Earth's resources.* Forest Row, U.K.: Clairview Books.

Jackson, T. (2009). *Prosperity without growth.* London: Earthscan.

Jevons, W. S. (1865/1965). *The coal question; an inquiry concerning the progress of the Nation, and the probable exhaustion of our coal-mines* (reprint, 3rd revised edition ed.). (A.W. Flux, Ed.,) New York: A.M. Kelly.

Lyle, J. T. (1994). *Regenerative design for sustainable development.* New York: Wiley.

Mehlman, P., (Writers) Gross, M., & (Director) Ackerman, A. (1996). Seinfeld: The shower head [Television series episode]. In *Seinfeld* (by L. David). New York: Castle Rock Entertainment.

Mims, C. (2013, November 13). *Forget LED bulbs—the future of interior lighting is lasers.* Retrieved Nov 18, 2013, from Quartz: http://qz.com/146761/forget-led-bulbs-the-future-of-interior-lighting-is-lasers/

Mokhtarian, P. (2009). If telecommunication is such a good substitute for travel, why does congestion continue to get worse? *Transportation Letters, 1,* 1–17.

Nordhaus, W. D. (1996). Do real-output and real-wage measures capture reality? The history of lighting suggests not. In T. F. Bresnahan & R. J. Gordon (Eds.), *The economics of new good* (pp. 27–70). Chicago: University of Chicago Press.

Shelley, C. (2016). Ideology in bio-inspired design. In L. Magnani & C. Casadio (Eds.), *Model-based reasoning in science and technology: Logical, epistemological, and cognitive issues* (Vol. 27, pp. 43–56). Berlin: Springer.

The Economist. (2007, June 18). *Burning bodies better.* Retrieved July 11, 2016, from The Economist: http://www.economist.com/node/9354014

Tsao, J. Y. (2010). Solid-state lighting: An energy-economics perspective. *Journal of Physics. D. Applied Physics, 43*(35), 345001.

Wolfe, A. (2015, November 20). *How washing machines could use a lot less water.* Retrieved November 23, 2015, from The Atlantic Citylab: http://www.citylab.com/tech/2015/11/how-washing-machines-could-use-a-lot-less-water/416961/

Videos

Video materials, both long and short, can enhance the educational value of the material presented in the preceding chapters. They may be used by readers to fill out the material contained in each chapter. They may be used by educators to provide instructional experiences that cannot be obtained through reading. In this section, instructive video material is listed for each chapter in the book.

Some materials listed directly concern items mentioned in the text. Other videos are of broader relevance to each topic. The latter are included for general reference.

Good Design

Garcia, P., (Director) & Alvarez, E. (Producer). (2005). *Bauhaus: Less is more.* United States: Film Media Group.

Kirby, T. (Director and Producer). (2010). *The genius of design: Ghosts in the machine.* United Kingdon: BBC.

Tuvie. (2015, 6 May). *Bruno Smart Trashcan by Poubelle LLC.* [Video file]. Retrieved from: https://www.youtube.com/watch?v=GUiEt6fOOJ0.

Umbra. (2011, 21 January). *Garbino can by Umbra featured on Design DNA.* [Video file]. Retrieved from https://www.youtube.com/watch?v=GUiEt6fOOJ0.

Social Psychology

IG St. Pauli. (2015, 2 March). *St. Pauli pinkelt zurück // St. Pauli Peeback.* [Video file]. Retrieved from https://www.youtube.com/watch?v=uoN5EteWCH8.

Culture

Black Berry. (2010, 14 June). *Jaipur foot.* [Video file]. Retrieved from https://www.youtube.com/watch?v=T2XhHxvE-Es.

© Springer International Publishing AG 2017
C. Shelley, *Design and Society: Social Issues in Technological Design*,
Studies in Applied Philosophy, Epistemology and Rational Ethics 36,
DOI 10.1007/978-3-319-52515-0

Fettig, T. (Director), & Westrate, E. (Producer). (2006). *Adaptive reuse in the Netherlands*, episode 4 of *Design e2: The economies of being environmentally conscious*. United States: McIntyre Media.

Harbury, M. (Director), & Swing, C. (Producer). (2002). *Peanuts*. United States: Bullfrog Films.

Hustwit, G. (Director and Producer). (2007). *Helvetica*. United States: Plexifilm.

Kirby, T. (Director and Producer). (2010). *The genius of design: Designs for living*. United Kingdom: BBC.

Rajosh. (2011, 3 May). *Fjaertzenpiip—The world's most expensive toilet*. [Video file]. Retrieved from https://www.youtube.com/watch?v=dQpXoX-ZyB0.

Rodriguez, C. (Director), & Gallo, J. S. (Producer). *Living on water*. United States: Forward in Time.

Schirrman, D., & Kendall, A.-C. (Directors). (2006). *Design—volume 1* and *Design —volume 2*. France: Facets Multimedia.

Style

Jacques, A. (Writer). (1996). *The way we dress: The meaning of fashion*. United States: Learning Seed.

Protess, D. (Producer and Writer). (2013). *10 buildings that changed America*. United States: PBS.

Schrank, J., Phipps, R., & Lombardo, J. (Writers). (2009). *Reading blue jeans: Clothing and culture*. United States: Learning Seed.

Social Agendas

Breitbart, E. (Director and Producer). (1982). *Clockwork*. United States: California Newsreel.

Clark, T., & Ryan, J. (Producers). (2000). *In the mind of the architect, Part 1: Keeping the faith*. Australia: Australian Broadcasting Corporation.

coffeekid99 (2007, 30 April). *The crying indian—full commercial: Keep America Beautiful*. [Video file]. Retrieved from https://www.youtube.com/watch?v= j7OHG7tHrNM.

Hablesreiter, M., & Stummerer, S. (Directors). Geyrhalter, N., Kitzberger, M., & Glaser, M. (Producers). (2009). *Food design*. United States: Icarus Films.

Hughes, R. (Writer). (1980). *The shock of the new, episode 4: Trouble in Utopia*. United Kingdom: BBC.

Rolighetsteorin (2009, 7 October). *The world's deepest bin—Thefuntheory. com/Rolighetsteorin.se*. [Video file]. Retrieved from https://www.youtube.com/ watch?v=cbEKAwCoCKw.

thwartd. (2011, 4 December). *Ikea Lamp TV commercial.* [Video file]. Retrieved from https://www.youtube.com/watch?v=Nix6tC3vvjs.

Vaughan, K. (Director and Producer). (2008). *The museum.* Canada: National Film Board.

Activism

Creadon, P. (Director), O'Malley, C., & Baer, N. (Directors). (2013). *If you build it.* United States: Long Shot Factory.

Fettig, T. (Director), & Westrate, E. (Producer). (2006). *Green for all*, episode 2 of *Design e2: The economies of being environmentally conscious.* United States: McIntyre Media.

Hewitt, C (2010, 23 April). *Why design now?: The learning landscape.* [Video file]. Retrieved from https://www.youtube.com/watch?v=empdZU-1_i8.

Poptech (2009, 2 November). *PopTech 2009 social innovation fellow Emily Pilloton.* [Video file]. Retrieved from https://vimeo.com/7393447.

OutofPoverty (2008, 13 February). *Out of poverty: Paul Polak on practical problem solving.* [Video file]. Retrieved from https://www.youtube.com/watch?v= kSEGN-EJJho.

Roberts, J. (2016, 3 June). *MOM: The inflatable incubator CBS News report.* [Video file]. Retrieved from https://www.youtube.com/watch?v=SKL-6Tj6PWc.

Schultz, C. (Director and Producer). (2002). *The Rural studio.* United States: BluePrint Productions.

TED (2010,18 October). *Jessica Jackley: Poverty, money—and love.* [Video file]. Retrieved from https://www.youtube.com/watch?v=Cqj0sgrNL10.

TED (2013, 19 December). *Krista Donaldson: The $80 prosthetic knee that's changing lives.* [Video file]. Retrieved from https://www.youtube.com/watch?v= LIy2oVJtJsA.

TEDxSacramento (2010, 23 June). *Robyn Waxman—FARM.* [Video file]. Retrieved from https://www.youtube.com/watch?v=1knKJv3bgIw.

Social Spaces

Dalsgaard, A. (Director), & Sørensen, S. B. (Producer). (2012). *The human scale.* Canada: Mongrel Media.

Fettig, T. (Director), & Westrate, E. (Producer). (2006). *Bogotà*, episode 2 of *Design e2: The Economies Of Being Environmentally Conscious.* United States: McIntyre Media.

Hustwit, G. (Director and Producer). (2011). *Urbanized.* United States: Plexifilm.

Klodawsky, H. (Director), F, I., & Martin-Gousset, L. (Producers). (2009). *Malls R Us.* United States: Icarus Films.

Mansfield, J. M. (Producer). (1973). *The writing on the wall.* United Kingdom: British Broadcasting Corporation.

Nawaz, Z. (Director), & MacDonald, J. (Producer). (2005). *Me & The Mosque*. Canada: National Film Board.

Shin (2015, 8 January). *Turn to the future*. [Video file]. Retrieved from https://vimeo.com/116303272.

Vohra, P. (Director). (2006). *Q2P: Toilets And The City*. India: V Tape.

Weyman, B. (Director), Weyman, B., & Allder, M. (Producers). (1994). *Return to Regent Park*. Canada: National Film Board.

Whyte, W. H. (Director and Producer). (1988). *The social life of small urban spaces*. United States: Municipal Art Society of New York.

Wiland, H., & Bell, D. (Directors and Producers). (2012). *Designing healthy communities, episode 4: Searching for Shangri-La*. United States: Media & Policy Center Foundation.

Risk

Dezeen (2015, 28 January). *Jaguar's Bike Sense could tap drivers on the shoulder to alert them of nearby cyclists*. [Video file]. Retrieved from https://www.youtube.com/watch?v=OXhlHGAjNxE.

Electrek.co (2016, 23 May). *Tesla Model S driver caught sleeping at the wheel while on Autopilot*. [Video file]. Retrieved from https://www.youtube.com/watch?v=sXls4cdEv7c.

TED (2011, 27 April). *Bruce Schneier: The security mirage*. [Video file]. Retrieved from https://www.youtube.com/watch?v=wQJC2MMB8nA.

Fairness

KMVT (2007, 14 May). *Tech closeup: Shot spotter*. [Video file]. Retrieved from https://www.youtube.com/watch?v=V28UMrWGARk.

NewsFortheLocals (2013, 31 January). *Avoid the Black People App*. [Video file]. Retrieved from https://www.youtube.com/watch?v=5kJqEMgdYiE.

Progress

de Rouvre, C.-A., & Scemla, J. (Directors). (2009). *Welcome to the Nanoworld—Episode 4: Nanoworlds and maxi-fears*. France: McNabb Connolly.

Mair, L. (Producer), & Kaldor, L. (Director). (2013). *Shattered ground*. Canada: Zoot Pictures.

Palfreman, J. (Director and producer). (2010). *Nuclear aftershocks: Has the world turned its back on nuclear energy?* United States: PBS.

Thompson, J., & Thompson, B. (Directors and producers). (2010). *Playing God with planet earth*. Canada: Lightship Entertainment Inc.

Winch, J. (Director and producer). (2011). *Bending the rails*. Canada: Moving Images.

Sustainability

Fettig, T. (Director), & Westrate, E. (Producer). (2006). *Deeper shades of green, episode 6 of Design e2: The economies of being environmentally conscious*. United States: McIntyre Media.

Feydel, S. (Director and producer). (2006). *Oceans of plastic*. United States: Landmark Media.

Lake, S. (Director), & Merrifield, A. (Producer). (2012). *Drying for freedom*. United States: Video Project.

Neighborgoods (2011,11 February). *How neighborgoods works*. [Video file]. Retrieved from https://vimeo.com/19846300.

Northcutt, P. (Director). (2005). *The ecological footprint: Accounting for a small planet*. United States: Northcutt Productions.

Index

© Springer International Publishing AG 2017
C. Shelley, *Design and Society: Social Issues in Technological Design*,
Studies in Applied Philosophy, Epistemology and Rational Ethics 36,
DOI 10.1007/978-3-319-52515-0